AF552734

ENCYCLOPAEDIA OF DEVELOPMENTAL BIOLOGY-V

DEVELOPMENT OF CHICK

By

Manju Yadav

Lecturer

Department of Zoology

M.M.H. College

Ghaziabad (U.P.)

(India)

DISCOVERY PUBLISHING HOUSE PVT. LTD.

NEW DELHI-110 002

First Published-2008

ISBN 978-81-8356-301-7

Published by

DISCOVERY PUBLISHING HOUSE PVT. LTD.

4831/24, Ansari Road, Prahlad Street,
Darya Ganj, New Delhi-110002 (India)
Phone: 23279245 • Fax: 91-11-23253475
E-mail: dphbooks@rediffmail.com
dphtemp@indiatimes.com

Printed at:

Sachin Printers, Delhi

Preface

The present title **Development of Chick** is an important link between almost all biological sciences, and is assuming more and more importance in fields such as medicine and agriculture. Molecular biology has began fulfilling its long anticipated role of linking genetics and embryology and many questions of embryology that had lain dormant for decades have been taken up with new molecular tools and strategies. While our knowledge of the molecular aspects of development has so drastically increased during the time, our applications of developmental phenomenon in evolution and ecology has also expanded. It is becoming impossible to study any area of biology without some background in developmental biology.

This title integrates the descriptive, experimental and biochemical approaches into a conceptual framework for the analysis of development. All important points are illustrated diagrammatically. It is well balanced, completely accessible presentation of principles and application of embryology. In order to create a book which is written within the economy as well as the physical group of the student, only the main essentials of each topic have been presented. It does this with the aid of careful selected examples—some recent and other classic of the field and with numerous illustrations. The aim is to enthuse the reader with this active and exciting area of research and to lay a solid foundation on which further study of its various facts may be based.

To make the work more comprehensive and informative, the author has consulted many authoritative books, research journals, abstracts, monographs etc. He is grateful to all those great scholars whose work are cited or substantially reproduced.

There can be no claim to originality except in the manner of treatment and much of the information has been obtained from the books and scientific journals available in the different libraries.

The author expresses his thanks to his friends and colleagues whose continue inspirations have initiated him to bring out this book.

The author expresses his gratitude to Mr. Wasan and staff of M/s Discovery Publishing House for their whole hearted co-operation in the publication of this book.

Author

CONTENTS

1

Introduction

EMBRYOLOGY

Every one of the higher animals starts life as a single cell—the fertilized ovum. This fertilized ovum, as its technical name *zygote* implies, has a dual origin. It is formed by the fusion of a germ cell from the male parent with one from the female parent. The union of two such sex cells to form a zygote constitutes the process of *fertilization* and initiates the life of a new individual. Embryology is the study of the growth and differentiation undergone by an organism in the course of its development from a single fertilized egg cell into a highly complex and independent living being like its parents.

As a study embryology offers more than the mere acquisition of an interesting array of facts. It gives one an understanding of some of the ways of life. We are all egoists, to a certain extent at least, and anything that touches the matter of our own whence and whither is of absorbing interest. The processes by which a fish, or an alligator, or a chick grows from a single fertilized egg cell to its fully elaborated adult structure are fundamentally the same as those involved in our own development. And these growth processes hold for us something definite and tangible in answer to that ever recurring question, "Whence do we come, and how?"

Embryology is also an important source of evidence as to the path followed by evolution. It tells us in one short, uninterrupted story how each individual grows into an adult. We can see this process going on under our very eyes. And we know that the story of individual development sketches for us in outline the evolutionary changes of our forbears. For the law of biogenesis or recapitulation is that *every living thing, in its individual development, passes through a series of constructive stages like those in the evolutionary development o f the race to which it belongs.*

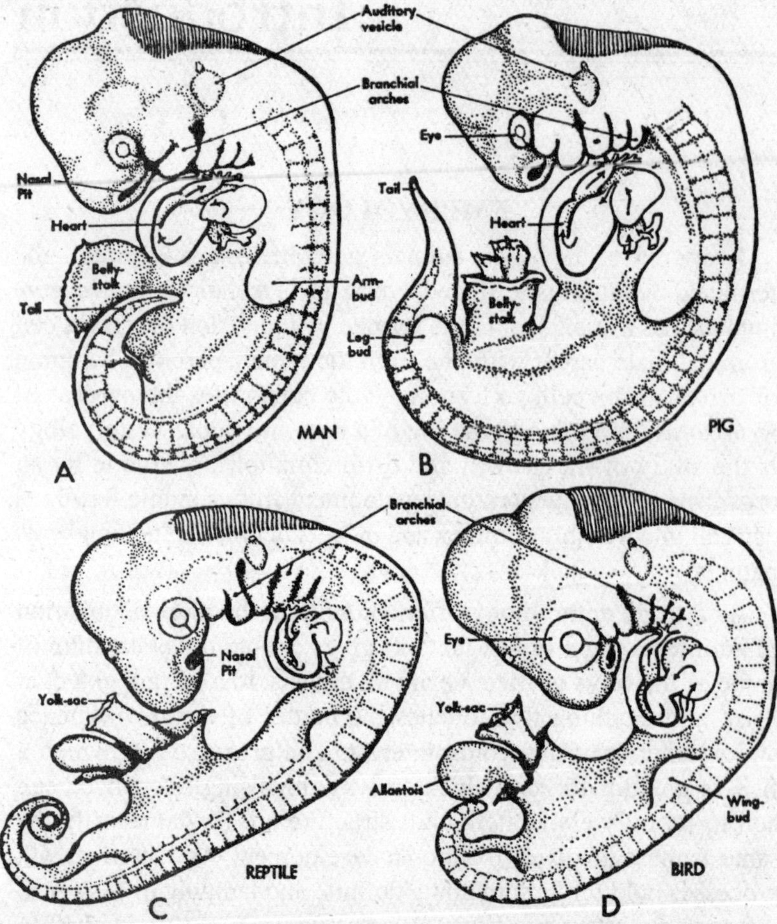

Figure 1.1 : Embryos of (A) man, (B) pig, (C) reptile, and (D), bird at corresponding developmental stages. The striking resemblance of the embryos to one another is indivative of the fundamental similarity of the processes involved in their development.

This means that there is but one main way to upbuild a given kind of organism, and that every individual must do it in essentially the same way his ancestors did. But to-day the individual can do this much more quickly and economically than its ancestors, because it does not have to find out by long and costly experimenting just how to do it. Essentially the right ways and means of doing it "automatically" are provided for it by its antecedents. At every step in the process it is using something which Nature, out of all the efforts of the past, has provided for it. These ways and means constitute the Heritages of individual life, and range all the way from ancestral germinal materials, stored up foods, and other parental provisions, to the environmental physical and social provisions in which the growing organism is placed.

More tangible perhaps than the perspective embryology gives on the life of today and on its evolutionary history in the past, is its direct help in the study of anatomy. Not until the student becomes enmeshed in a maze of structural details does he ,realize his imperative need of a knowledge of how and why adult conditions became as they are. For only this knowledge will lead him beyond blind memorizing to comprehension. And when he delves still deeper into anatomy, he begins to encounter puzzling variations from the normal body architecture. Sometimes these are merely minor anomalies which do not materially affect the functional fitness of the individual; sometimes they are extensive malformations which render continued life precarious or even altogether impossible. Our understanding of such conditions, and what future hope there may be of reducing the frequency with which they occur, can come only through extending our knowledge of genetics and embryology.

The only method of attaining a comprehensive understanding of embryological processes is through the study and comparison of development in various animals. Many phases of the development of any specific organism can be interpreted only through a knowledge of corresponding processes in other organisms. The beginning student, however, can most readily and with least risk of confusion acquire his knowledge of embryology through intensive study of one form at a time. Building on the familiarity with fundamental processes of development thus acquired, he may later broaden his horizon by the comparative study of a variety of forms.

THE CHICK AS LABORATORY MATERIAL

The chick is one of the most satisfactory animals on which student laboratory work in embryology may be based. Chick embryos in a proper state of preservation and of the stages desired can readily be secured and prepared for study. Used as the only laboratory material in a brief course, they afford a basis for understanding the early differentiation of the organ systems and the fundamental processes of body formation common to all groups of vertebrates. In more extended courses where several forms are taken up, the chick serves at once as a type for the development characteristic of the large-yolked eggs of birds and reptiles and as an intermediate form bridging the gap between the simpler processes of development in fishes and amphibia on the one hand and the more complex processes in mammals on the other. In medical school courses where a knowledge of human embryology is the end in view, the chick not only makes a good stepping stone to the understanding of mammalian embryology, but. also provides material for the study of early developmental processes not readily demonstrable in human material.

PLAN AND SCOPE OF THIS BOOK

This book on the development of the chick has been written for those who are beginning the study of embryology and has accordingly been kept as brief and uncomplicated as possible. Nevertheless it is assumed that the beginner in embryology will not be without a certain background of zoological knowledge and training. He may reasonably be expected to be familiar with the fundamental facts of evolution and heredity, the structure of cells and their methods of division, the nature of the various types of tissues, and the more general phases of the morphology of vertebrates. It therefore seems unnecessary to include here any preliminary discussion of these phenomena.

Because I am convinced that it is not a logical approach for a beginner, this book does not emphasize the comparative viewpoint.

Nevertheless, where the comparative background for some particular phase of chick development seemed to promise to make it more intelligible, I have not hesitated to bring it in. This has been done especially freely in connection with cleavage and gastrulation in the early parts of the book, and in connection with the cardiovascular and excretory systems in the final chanter. It is hoped that these excursions, instead of rendering the student's task more onerous, will lighten it by making these phases of development more readily understandable.

It has seemed wise to take essentially the same standpoint with reference to experimental embryology. This rapidly growing and important field is one which can be handled best in an advanced course, building on a background such as this book aims to provide for beginners. Accordingly, no attempt has been made to deal with it systematically. But, as with comparative embryology, when the experimental approach seemed to offer particular help in interpreting some phase of development, it has been freely drawn upon. References for collateral reading on these and other phases of the subject are given in the bibliography.

METHODS OF STUDY

Like other sciences embryology demands first of all accurate observation. It differs considerably, however, from such a science as adult anatomy where the objects studied are relatively constant and their component parts are not subject to rapid changes in their interrelations. During development, structural conditions within the embryo are constantly changing. Each phase of development presents a new complex of conditions and new problems.

Solution of the problems presented in any given stage of development depends upon a knowledge of the stages which precede it. To comprehend the embryology of an organism one must, therefore, start at the beginning of its development and follow in their natural order the changes which occur. At the outset of his work the student must realize that proper sequence of study is essential and may not be disregarded. A knowledge of structural conditions in earlier stages than that at the moment under consideration and an appreciation of the trend of the developmental processes by which conditions at one stage become transmuted into different conditions in the next are direct and necessary factors in acquiring a real comprehension of the subject. Without them the story of embryology becomes incoherent, a mere jumble of confused impressions.

A knowledge of the phenomena of development is ordinarily acquired by studying a series of embryos at various stages of advancement. Each stage should be studied not so much for itself, as for the evidence it affords of the progress of development. In the study of embryology it does not suffice to acquire merely a series of "still pictures" of various structures, however accurate these pictures may be. The study demands a constant application of correlative reasoning and an appreciation of the mechanical factors involved in the relations to each other of various structures within the embryo, and in the relation of the embryo as a whole to its environment. In order really to comprehend the embryologic significance of a structure one must know not only its relations within the embryo being studied at the time, but also the manner in which it has been derived and the nature of the changes by which it is progressing toward adult conditions. To get absolutely the whole story it is obvious that one would have to study a series of embryos with infinitely small intervals between them. Nevertheless the fundamental steps in the process may be grasped from a much less extensive series. The fewer the stages studied, however, the more careful must one be to keep in mind the continuity of the. processes and to think out the changes by which one stage leads to the next.

The outstanding idea to be kept in mind by the student beginning the study of embryology is that the development of an individual is a process and that this process is continuous. The conditions he sees in embryos of various stages are of importance chiefly because they serve as evidence of events in the process of development at various intervals in its continuity, as historical events are evidence of the progress of a nation. Just as historical events are led up to by preparatory occurrences and followed by results which in turn affect later events, so in embryology events in development are presaged by preliminary changes and when consummated affect in turn later steps in the process.

In certain respects the laboratory study of embryological material involves methods of work for which courses in general zoology do not entirely prepare the student. Some general suggestions as to methods of procedure are, therefore, not out of place.

In dissecting gross material it is not unduly difficult to appreciate the complete relationships of a structure. The nature of embryological material, however, introduces new problems. Embryos of the age when the establishment of the various organ systems and processes of body formation are being initiated are too small to admit of successful dissection, but not sufficiently small to permit of the satisfactory microscopical study of an intact embryo, except for its more general organization. To study embryos of this stage with any degree of thoroughness, they must be cut into sections which are sufficiently thin to allow effective use of the microscope to ascertain cellular organization and detailed structural relationships. In preparing such material the entire embryo is cut into sections which are mounted on slides in the order in which they were cut. A sectional view of any region of the embryo is then available for study.

While sections readily yield accurate information about local regions, it is extremely difficult to construct a mental picture of any whole organism from a study of serial sections alone. For this reason it is necessary to work first on entire embryos which have been prepared by staining and clearing so they may be studied as transparent objects. From such preparations it is possible to map out the configuration of the body and the location and **extent** of the more conspicuous internal organs. In this work the fact that embryos have three dimensions must be kept constantly in mind, and, by careful focusing, the depth at which a structure lies must be determined as well as its apparent position in surface view. Although, conventionally, entire chick embryos are usually represented in dorsal view, much additional information may be gained by following a study of the dorsal, with a study of the ventral aspect. Unless the preliminary study of entire embryos is carefully and thoughtfully carried out, the study of sections is likely to yield only confusion.

In studying a section from a series it is necessary first of all to determine the location in which it was cut through the embryo. The plane of the section under consideration, as well as the region of the embryo through which it passes, should be ascertained by comparing it with an entire embryo of the same age as that from which the section was cut. Only when the location of a section is known precisely can the structures appearing in it be correlated with the organization of the embryo as a whole. Probably nothing in the study of embryology causes students more difficulties than neglect to locate sections accurately, with the consequent failure to appreciate the relationships of the structures seen in them. Too great emphasis cannot be laid on the vital importance of fitting the structures shown by sections properly into the general scheme of organization as it appears in whole-mounts.

It must by no means be inferred that the possibilities ot the wholemounts have been exhausted by the preliminary study accorded them before taking up the work on sections. Further and more careful study of entire embryos should constantly accompany

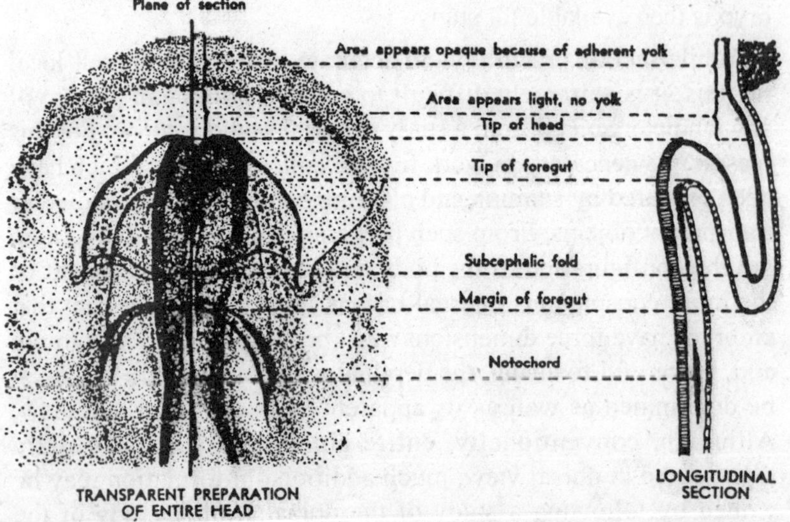

Figure 1.2 : Relation of longitudinal section of the embryonic head to the picture presented by a head of the same age mounted entire as a transparent preparation.

the study of serial sections. Many details which in the initial observation of the whole-mount were inconspicuous or abstruse will become significant in the light of the more exact information yielded by the sections. Relative levels of closely related structures

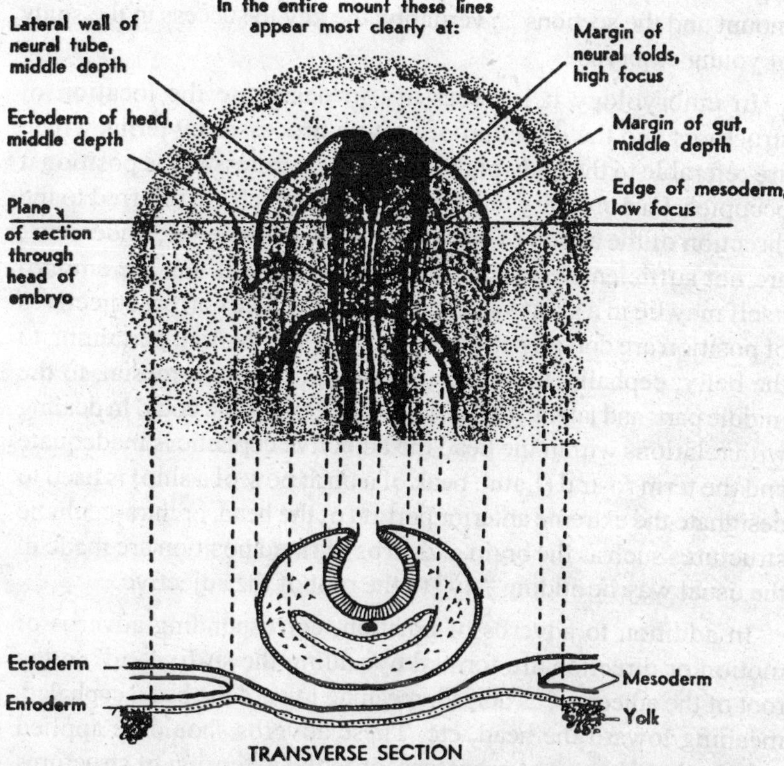

Figure 1.3 : Relation of transverse section of the embryonic head to the picture presented by an entire head of the same age viewed as a transparent preparation. The two drawings may be brought into closer relation by looking at them with the top of the page tilted downward. This and the preceding figure show the aid that sections can be in interpreting entire embryos and, at the same time, the way in which the greater perspective lent by study of entire embryos increases the significance of sectional pictures. Note especially that neither the transverse nor the longitudinal section alone adequately interprets all the features of the entire mount. An embryo is a three-dimensional object and must be studied with that fact in mind. These figures show, also, how largely the appearance of lines or bands seen in studying transparent mounts of entire embryos is due to looking edgewise through a folded layer.

which are particularly difficult to work out by focusing through a wholemount are unequivocally shown by sections. Conversely, there is nothing that will help a beginner as much in the interpretation of serial sections as constantly relating them to the organization of the entire embryo. Correlative study of the whole-mount and the sections is, veritably, the key to success in the study of young embryos.

In embryology it is necessary to designate the location of structures and the direction of growth processes by terms which are referable to the body of the embryo regardless of the position it occupies. Our ordinary terms of location which are referred to the direction of the action of gravity, such as above, over, under, etc., are not sufficiently accurate because of the fact that the embryo itself may lie in a great variety of positions. The correct adjectives of position are dorsal, pertaining to the back; ventral, pertaining to the belly; cephalic, to the head; caudal, to the tail; mesial, to the middle part; and lateral, to the side of the embryonic body. In dealing with relations within the head the adjective cephalic is inadequate and the term rostral (Latin, beak of a bird; bow of a ship) is used to designate the extreme anterior portion of the head, or intra-cephalic structures such as the brain. Adverbs of fixed position are made in the usual way be adding "-ly" to the root of the adjective.

In addition to adverbs of position, corresponding adverbs of motion or direction are formed by adding the suffix "-ad" to the root of the adjective, as dorsad, meaning toward the back, cephalad, meaning toward the head, etc. These adverbs should be applied only to the progress of processes, or to the extension of structures toward the part indicated by their root. Thus, for example, we should say that the developing eye of an embryo was lo cated in the lateral wall of the forebrain or that the forebrain was in the cephalic part of the embryo; but if we wished to express the idea that as the eye increased in size it moved farther to the side we should say it grew laterad. Cultivation of the use of correct and definite terms of position and direction in dealing with embryological processes will greatly aid accurate thinking and clear understanding.

2

The Gametes and Fertilization

CONTINUITY OF GERM PLASM

The reproductive cells which unite to initiate the development of a new individual are known as *gametes.* In the case of all the higher organisms the gametes are of two types produced by two sexually differentiated individuals, the small, actively motile gametes from the male being called *spermatozoa* or *spermia,* and the larger, food-laden gametes formed within the female being termed *ova.* The gametes and the cells which give rise to them are said to constitute the individual's *germ plasm.* The cells which take no direct part in the production of gametes are called somatic cells. In antithesis to the germ plasm, the somatic cells collectively are said to constitute the *somatoplasm.*

The germ plasm is of paramount interest, not only to the biologist, but to all thinking persons, because while the somatic cells cease to exist with the death of the individual whose body they constitute, the germ plasm may live on indefinitely in succeeding generations. In the higher vertebrates where sexual reproduction is the rule, the germ plasm of a single individual cannot survive by itself. There must be successful union of a male with a female sex cell. These two conjugating gametes alone pass on the entire hereditary dowry of the species. It is not easy to realize fully all that is implied by this simple statement as to the continuity of the germ plasm. It may make it more vivid if we apply it specifically to ourselves and say that the future of the human race depends on the germ plasm held in trust within the bodies of the individuals now living. Fortunately, very early in the life of an individual, the germ plasm is segregated in the gonads and is not subject to most of the vicissitudes and the diseases from which the somatic cells suffer. But the germ plasm, even though it is not directly affected, may nevertheless suffer indirectly because of a poor environment forced upon it by an unhealthy body.

The quality of the germ plasm of any one individual, while of great importance, is only half the story. Of equal significance is the nature of the combination of, germ plasm which occurs in each generation when the male and female gametes fuse. As surely as either gamete brings into the new combination defective germ plasm, so surely will both the body and the germ plasm of the new individual suffer therefrom.

HISTORY OF SEX CELLS WITHIN THE PARENT BODY

It is, therefore, of fundamental interest to go back of the production of gametes in the sexually mature individual so we may see all the steps in the preservation and transmission of the racial heritage. Obviously the germ plasm of any individual must have come to it from its parents by way of the ovum and the spermium. But one wants to know how and where the germ plasm was cared for by the parents; when in their life history it first became possible to recognize it as distinct from the somatoplasm; when and how it became segregated in the gonads; and what it was doing during the long period before sexual maturity.

There is still much of the very early history of the germ plasm which is but imperfectly known. Yet the cells which are destined to give rise to the gametes are definitely recognizable at a surprisingly early stage in development. Even before it is possible to tell whether an embryo is destined to become a male or a female, certain large cells, different from their neighbors, can be recognized as the *primordial sex cells*. In other words we know that they constitute the germ plasm as it exists in an embryo of that age.

The primordial sex cells are first easily identifiable when the gonads are just taking shape as definite organs. Until recently it was believed that, in the higher vertebrates, they could not be recognized as potential germ cells any earlier in their history. Recently much more detailed investigations have been made which seem to indicate that the sex cells can be recognized even prior to their appearance in the newly formed gonad. According to these studies the primordial germ cells become recognizably differentiated in the wall of the yolk-sac, and migrate thence to become established in the growing gonads. How much farther back toward the fertilized ovum the lineage of the primordial sex cells may in the future be traced, it would be unwise to predict. Even now, in some of the invertebrates, it is believed that the single cell, which gives rise to all the sex cells can be identified when the embryo is so young that it is no more than a minute ball of cells.

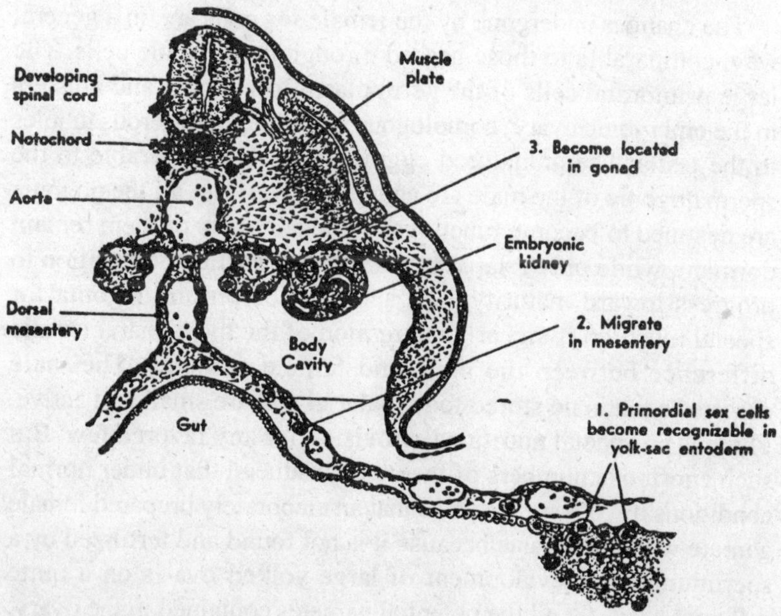

Figure 2.2 : Schematic section through the midbody region of a young embryo, illustrating the manner in which the primordial germ cells are believed to originate in the yolk-sac entoderm and migrate thence to the developing gonad.

The upper part of the chart summarizes graphically the conception of the early separation, from the somatic cells, of certain cells which are destined to give rise to the gametes. Since this process occurs before sexual differentiation, we may take it as a common starting point for the germ-cell lineage of either sex. The lower part of the chart outlines separately for each sex the later history of the germ cells. Sexual differentiation of the embryo begins to be apparent shortly after the primordial germ cells are established in the gonads. If the individual is to become a male, the gonads differentiate into testes. During early embryonic life the primordial sex cells become organized within the testis in the form of tortuous tubules called the *seminiferous tubules.* During the period of body growth which precedes sexual maturity, the future gamete-producing cells or *spermatogonia* in these tubules remain quiescent. During the period of active sexual life, the spermatogonia multiply and groups of them are constantly undergoing the final changes which precede their liberation as mature gametes.

The changes undergone by the female sex cells are, in a general way, comparable to those passed through by the male cells. The large primordial cells of the germ plasm form cords and clusters in the embryonic ovary, homologous with the seminiferous tubules in the testis. The unmatured eggmother cells comparable to the spermatogonia of the male are called *oogonia.* Not all the oogonia are destined to become functional gametes. Many of them remain dormant, while only relatively few receive sufficient nutrition to progress toward maturity. The selection of certain oogonia for special nutrition is an early expression of the most characteristic difference between the male and female gametes. The male gametes contain no stored food material and are small and active. There is no special nutritional provision for any favored few. But such enormous numbers of them are produced that under normal conditions there is little chance that an elaborately prepared female gamete will go to waste because it is not found and fertilized by a spermium. The development of large-yolked ova is on a quite different basis. Of all the potential gametes contained in the ovary, relatively few are destined to be brought to maturity. But these few are richly endowed with stored food materials. The energy expended by the male on quantity production of small active gametes ensuring fertilization is, thus, in the female, devoted to laying up a supply of food which provides the necessary materials for the growth of the embryo which is to result from fertilization.

SPERMATOGENESIS

The latter part of the history of the sex cells which occurs just before their liberation as mature gametes demands closer scrutiny. In the male these terminal changes are spoken of as constituting the process of spermatogenesis. The cells of the germ line which came to be located in the seminiferous tubules of the growing testis we have already learned to know by their technical name of spermatogonia. When a male animal becomes sexually mature these spermatogonia begin to produce gametes. They do not all become active at any one time, but periodically groups of spermatogonia begin to produce successive crops of mature spermia.

The process is much the same in all the higher vertebrates and it will be at once simpler and more profitable if we trace it in broad general lines rather than attempting to master its details in any one form. In sections of the testis, spermatogonia are found peripherally located in the walls of the seminiferous tubules. When a spermatogonium undergoes mitosis, either of the resulting cells may do one of two things. It may cease dividing for a time, and by growing to a size markedly larger than that of its parent, become differentiated as *a primary spermatocyte.* Or it may remain like its parent and take the place of cells which crowd toward the lumen of the tubule to become differentiated into spermatocytes. Some of the spermatogonial cells always remain thus in the peripheral part of the tubule and furnish a constant source of new cells ready for conversion into spermatocytes.

Once a cell has passed through the growth phase which so differentiates it that it is called a primary spermatocyte, its future history is definitely determined. As soon as its growth is completed, it undergoes two divisions in rapid succession. The first of these divisions results in the formation of two smaller cells called *secondary spermatocytes.* Each of these secondary spermat-

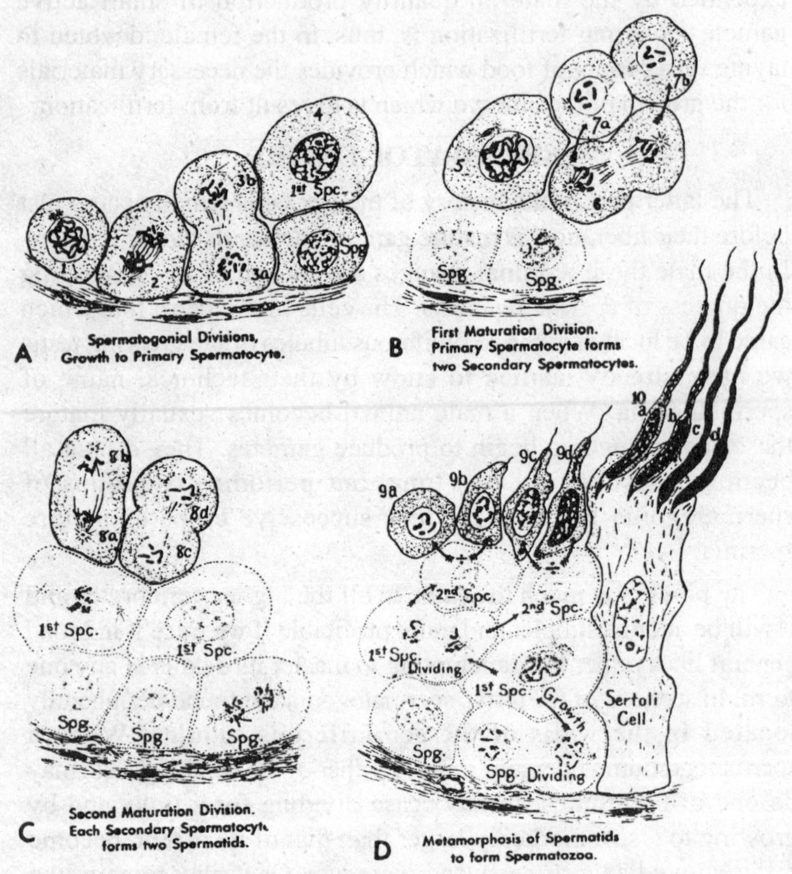

Figure 2.2 : Semischematic diagrams to show the main steps in the progress of spermatogenesis. In the wall of a mature seminiferous tubule, cells in all stages of spermatogenesis occur in such close association that it is not always easy to grasp their relations. To obviate this difficulty, these diagrams start with conditions in a young tubule just going into activity and follow the process a step at a time. The sequence of events is further indicated by consecutive numbering of the cells. In each part of the figure, cells undergoing the critical changes there emphasized are drawn in full detail, while cells not at the moment under consideration are "shadowed in" to make evident the accompanying changes in positional relations. For the sake of simplicity the species number of chromosomes is assumed to be eight.

ocytes, without any resting period which might allow the cells to grow to the size attained by their parents, promptly divides again. The four small cells thus produced from the primary spermatocyte by two successive cell divisions are known as *spermatids*. The two divisions are known as the *maturation (meiotic) divisions* both because they are the last divisons that these cells undergo and because they accomplish certain important internal changes preparatory to the part these cells are to play in fertilization. The nature of these changes, especially as they affect the chromosomal content of the cells, we shall consider in more detail in connection with the similar changes taking place in the maturation of the female sex cells.

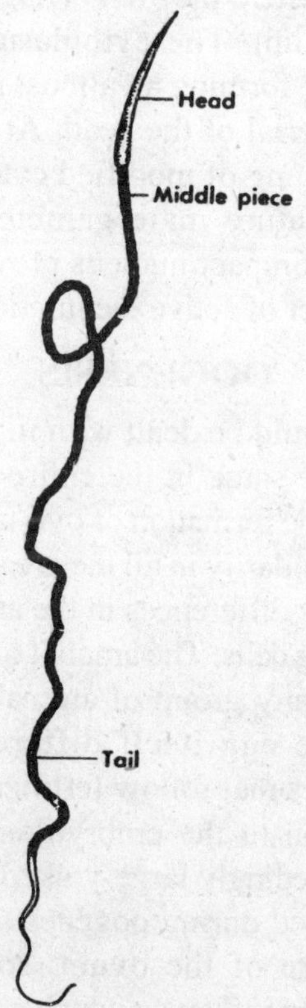

Figure 2.3 : Spermatozoon of the pigeon.

Although cell division ceases with the two maturation divisions, the spermatids still have radical changes to undergo before they are capable of carrying out their function. During this *metamorphosis of the spermatid* the nuclear material becomes very compact to form the bulk of the head of the spermium. The flagellum-like *tail* is formed by the development of a contractile fibril which first makes its appearance in the cytoplasm near the centrosome and then rapidly grows in length until it projects far beyond the original confines of the cell. As this tail filament grows it remains enveloped in a thin film of cytoplasm which eventually forms a membrane covering all but the extreme tip of the tail . The modified centrosomal apparatus becomes located in the so-called "neck region" where the "tail" (flagellum) is joined to the "head" of the spermium. The cytoplasm of the spermatid is reduced greatly in bulk forming an almost invisibly thin envelope about the nuclear material of the head. At the extreme tip of the head a cap-like thickening of modified cytoplasm constitutes the acrosome. Thus a mature male gamete is a cell consisting essentially of a very compact nucleus provided with a flagellum which gives it the power of active locomotion in a fluid medium.

OOGENESIS

Spermatogenesis could be dealt with in general terms because it is fundamentally the same in the entire vertebrate group. The processes involved in the formation of ova, although they also show a certain underlying similarity in all the great groups of vertebrates, are greatly modified by differences in the amount of food material stored as yolk in the egg cells. The amount of yolk characteristically present in the egg of any group of animals affects not only the manner in which the egg itself differentiates but also the devlopmental processes that follow fertilization. As we are going to devote our attention to the embryology of the chick which develops from an exceedingly large-yolked egg, we must see how this yolk is accumulated during oogenesis and how its presence modifies the structure of the ovum, so that we may better understand the profound influence which a large yolk-mass exerts on the early growth processes of the embryo.

The part of the hen's egg commonly known as the "yolk" is a single cell, the female sex cell or ovum. Its enormous size as compared with other cells is due to the food material it contains. This food material, or *deutoplasm,* destined to be used by the embryo in its growth, is gradually accumulated within the cytoplasm of the ovum before it is liberated from the ovary. Under the microscope it has the appearance of a viscid fluid in which are suspended granules and globules of various sizes. On chemical analysis yolk is found to contain proteins in various combinations, carbohydrates, and fatty substances in the form of lecithin and sterols as well as neutral fat. Vitamins A, B_1, B_2, D, and E are also present. As these stored materials increase in amount, the nucleus and the cytoplasm are forced toward the surface, so that eventually the deutoplasm comes to occupy nearly the entire cell.

The significance of the chick's liberal endowment of yolk can best be appreciated if one compares the course of its development with that of the frog. The yolk in a frog's egg, though considerable in amount, is not sufficient to carry the embryo through all the changes it must undergo before it attains a body organization like that of the parents. By the time all its yolk has been used the frog embryo has grown only to a larval form, commonly called a tadpole, which is still far short of adult structure. This tadpole is able to secure by its own activities sufficient food to complete its growth and to carry it through the remaining developmental steps which bring it out as a small frog. The chick has no such interrupted embryology. Its generous supply of yolk is sufficient to provide for a rapid and continuous development so that when it emerges from the shell it is already a miniature of its parents.

A section of the hen's ovary which includes several young ova and one ovum which is nearly ready for liberation. The very young ova lie deeply embedded in the substance of the ovary. As they accumulate more and more deutoplasm, they crowd toward the surface and finally project from it, maintaining their connection only by a constricted stalk of ovarian tissue. The protuberance containing the ovum is known as an *ovarian follicle.* The bulk of the ovum itself is made up of the yolk. Except in the neighborhood of the nucleus the active cytoplasm is but a thin film enveloping the yolk. About the nucleus a considerable mass of cytoplasm is aggregated. The region of the ovum containing the nucleus and the bulk of the active cytoplasm is known as the *animal pole* because this subsequently becomes the site of greatest protoplasmic activity. The region opposite the animal pole is called the *vegetative,* or *vegetal, pole* because, whereas material for growth is drawn from this region, it remains itself relatively less active.

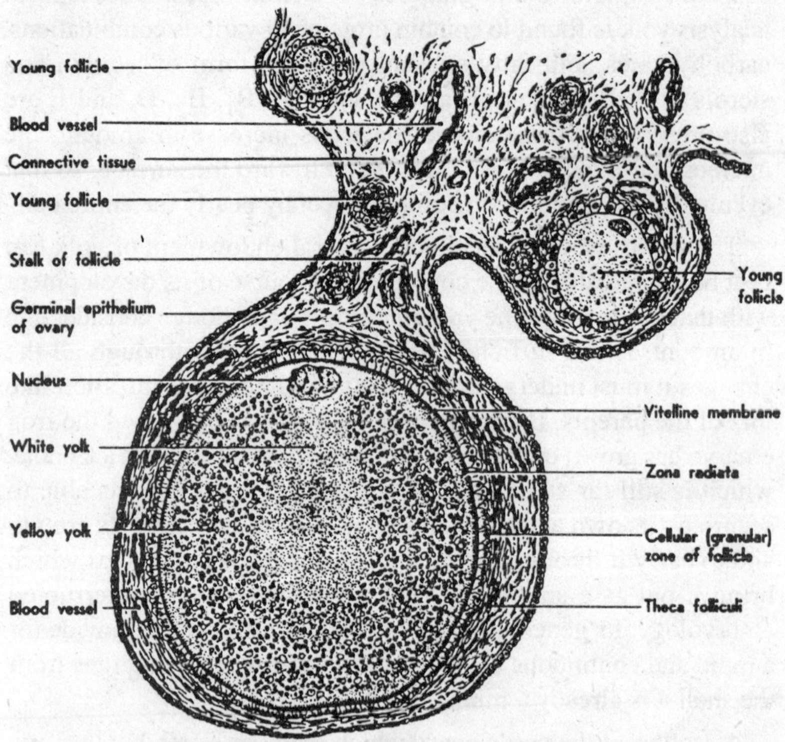

Figure 2.4 : Diagram showing the structure of a bird ovum still in the ovary. The section shows a follicle containing a nearly mature ovum, together with a small area of the adjacent ovarian tissue.

As is the case with the ova of many classes of vertebrates, the ovum of birds has a modified cell membrane called the *zona radiata.* This term was applied to it because, when viewed under high power with a light microscope, it seems to show delicate radial striations. Studied under the greater magnifications possible with an electron microscope this striated appearance is seen to be due to closely packed microvilli. The functional significance of such a structural arrangement is the great increase in membrane surface it affords, thereby enhancing the rate of the metabolic interchanges that can take place at the cell surface.

In a young ovum during the time it is rapidly accumulating deutoplasm the zona radiata is relatively conspicuous. As the amount of accumulated yolk increases, the striated appearance of the cell membrane becomes less evident, and it becomes definitely more robust. This secondarily thickened cell membrane of large-yolked ova is called the *vitelline membrane.*

Outside the vitelline membrane is an investment of small polygonal cells. These constitute the *cellular* or *granular zone* of the ovarian follicle. This cellular zone is in turn enclosed in a highly vascular layer of connective tissue known as the *theca folliculi.*

It should be borne in mind that the term "ovum," .used without qualification, does not carry a precise significance. It refers always to the female sex cell, but it must be specified whether one means a young ovum befores all the deutoplasm is accumulated, an ovum with its full complement of deutoplasm but unmatured, or a matured ovum ready for fertilization. The more exact terms which parallel those employed in designating the phases of spermatogenesis avoid this difficulty. The ovum during its period of growth and the accumulation of deutoplasm which we have just been considering is *a primary oöcyte,* developmentally homologous to the primary spermatocyte of the male. The cells of the granular zone of the ovarian follicle form a protective covering about the oögonium "chosen" for growth into a primary oöcyte and aid in transferring to it the nutriment supplied by the mother from the products of her digested food: This nutritive material is brought in through the blood vessels of the theca, absorbed by the follicular cells, and transferred by them to the growing primary oöcyte where it is elaborated into deutoplasm.

When the full allotment of deutoplasm has accumulated, the investing tissue of the theca is ruptured and the oöcyte is liberated from the ovary. Almost coincidently with *ovulation*, as the discharge of the ovum is called, the first maturation division occurs. In this the nuclear material of the primary oöcyte is halved between the two resulting cells but one of them gets practically all the cytoplasm and its contained food materials. Both of these unequal cells are *secondary oöcytes*, but only the one which received the entire dowry of deutoplasm has any chance of becoming functional. The small oöcyte containing practically no cytoplasm, because it was budded off from the "animal pole" of the ovum, is commonly called *a "polar body"* or *polocyte*. This polocyte drags along a discouraged existence for a time and may undergo an abortive second maturation division. But it is doomed to degenerate. Again in the second maturation division all the stored food material goes to one cell. The cell which receives no deutoplasm is called the *second polocyte* and is, like the first polocyte, destined for degeneration. The cell which has all the deutoplasm that might have been divided among four sister *oötids is* the mature gamete ready for fertilization.

If one reviews mentally the phenomena of oögenesis, one is impressed both by the underlying parallelism of oögenesis and spermatogenesis and by certain striking differences between the two processes. The early history of the germ plasm is the same in the two sexes, so too is the period of gametemother-cell multiplication, followed by the very definite sequence of a growth phase, which is in turn followed by two maturation divisions (Fig. 2.5). But each sex shows certain highly characteristic modifications of the process. Whereas in the male any spermatogonium may produce four mature gametes, in the female three potential oögonia are sacrificed in the more elaborate provision for a chosen one. For, as a result of the two maturation divisions, the oöcyte gives rise to three nonfunctional polar bodies and but one mature gamete which gets all the food accumulated by the primary oöcyte in its long growth period. The very liberality of the stored food material collected in the maturing ovum renders it inactive and passively dependent on surrounding structures for its transportation. In contrast, the concentration of nuclear material in their small, compact heads and the development of an elaborate propulsive

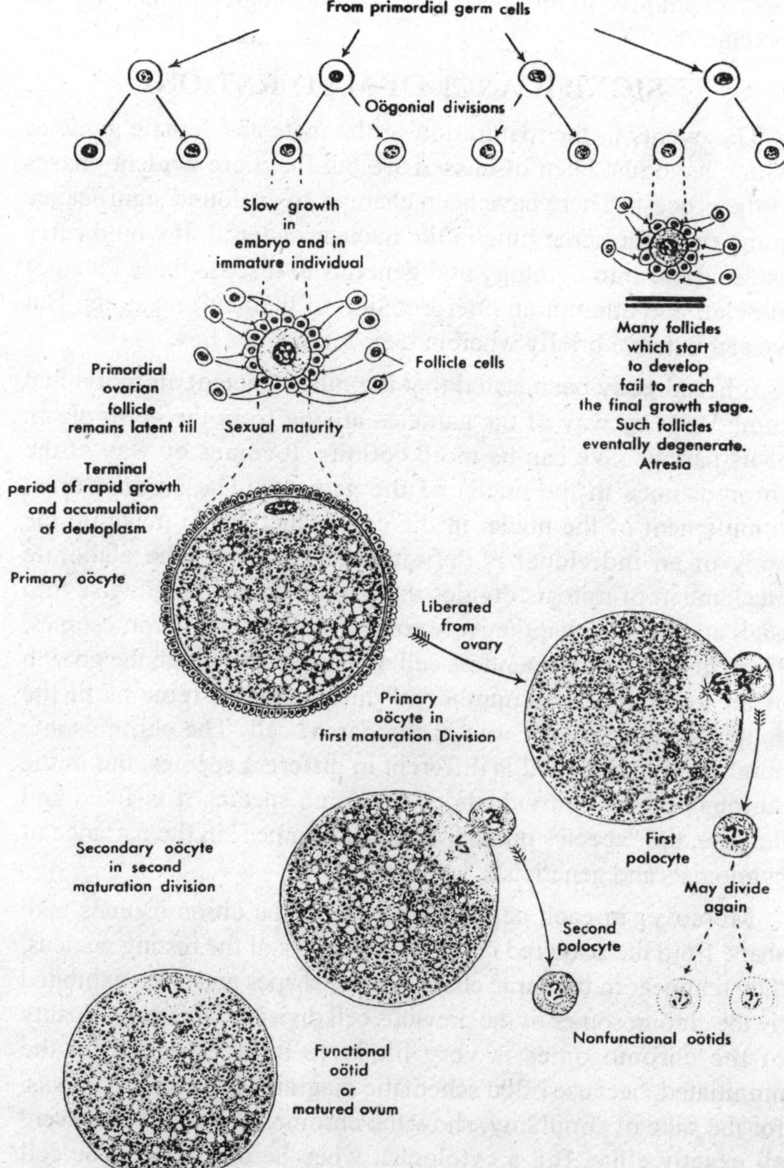

Figure 2.5 : Diagram to show the main steps in the growth and maturation of the ovum. For the sake of simplicity the species number of chromosomes is assumed to be eight.

mechanism give to the spermatozoa a high degree of independent motility.

SIGNIFICANCE OF MATURATION

The events in the maturation of the male and female gametes which have just been discussed are but the more evident phases of the process. There have been changes of profound significance going on at the same time in the nuclear material. It would carry us far afield into cytology and genetics to discuss these changes in detail and attempt an interpretation of their full meaning. But we can indicate briefly wherein their importance lies.

It has already been stated that the inheritance of an individual comes to it by way of the gametes arising from the germ plasm of its parents. We can be more definite. It comes by way of the chromosomes in the nuclei of the gametes. The chromosomal complement of the nuclei in the cells which go to make up the body of an individual is definite and constant. The elaborate mechanism of mitosis divides the chromosomes lengthwise into qualitatively and quantitatively equivalent daughter chromosomes. Thus, in each of the countless cell divisions involved in the growth of an individual, the number of chromosomes remains in the daughter cells what it was in the parent cell. The chromosome number so maintained is different in different species, but in the various cells of individuals of the same species it is fixed and definite, the "species number of chromosomes" in the parlance of cytologists and geneticists.

Moreover in each cell division, when the chromosomes take shape from the scattered chromatin granules of the resting nucleus, they reappear in the same characteristic shapes and sizes exhibited by the chromosomes of the previous cell division. This individuality of the chromosomes is very likely to be overlooked by the uninitiated, because often schematic diagrams illustrating mitosis, for the sake of simplicity, show the chromosomes as if they were all exactly alike. But a cytologist, when he has studied the cell structure of an animal intensively, will be able to tell us not only how many chromosomes we can count on finding at each mitosis but also how each chromosome will look. One he will characterize as long and bent, another as slender and straight, still another as short and plump, and so on. Furthermore on careful study of these chromosomes one finds that they are present in pairs, the members of which are similar in size and shape. The members of a vair are not usually located next to one another on the spindle of an ordinary somatic mitosis, but methodical comparison of their size and shape enables the cytologist to chart the chromosomes of a cell, similar pair, by similar pair.

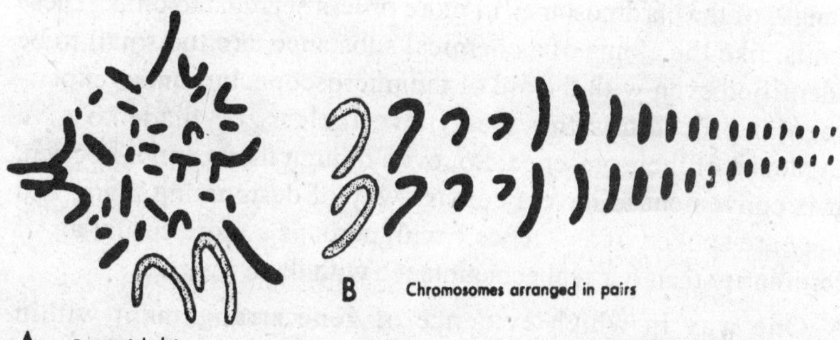

Figure 2.6 : The chromosomes of the fowl. The sex chromosomes are indicated by stippling. (A) Actual appearance (camera lucida drawing, polar view) of the chromosomes as they lie in the equatorial plane of a mitotic spindle just before the beginning of the metaphase splitting. (B) Diagram, to the same scale as A, showing how the chromosomes may be arranged in pairs, the members of which are similar to each other in size and shape.

The fact that the chromosomes thus occur in recognizable couples is of great significance to the geneticist in his study of the manner in which hereditary traits are passed on from one generation to the next. We are already familiar with the fact that all the cells of an animal's body have been formed by the division and redivision of the fertilized egg cell or zygote. We know that in each of these mitotic divisions each chromosome is split in half so that the daughter cells receive the same number of chromosomes that the parent cell contained. Obviously then, if a cell taken from a rooster's comb or anywhere else in his body showed 36 chromosomes, the zygote from which the rooster grew inevitably had 36 chromosomes. Now if we study the formation of the zygote, we find that of its full species number of chromosomes half were brought in by the male gamete and half by the female gamete.

And if we study the individual shapes and sizes of their chromosomes we find that each gamete contributed one member of each pair of chromosomes which characteristically-appear in the cells of the body in that species.

Much work has been done by geneticists and cytologists in an effort to link specific bodily traits with specific chromosomal structure. It is now quite generally believed that the material responsible for transmitting hereditary traits is strung out along the length of the chromosomes in more or less segregated units. These units, like the atoms of a chemical substance, are too small to be identified even with the aid of the microscope, but under experimental procedures they can, nevertheless, be made to give evidence of their presence. So, even though they cannot be seen, it is convenient to have a concise way of designating them, and they are spoken of as "genes," with perhaps a more casual air of familiarity than our real acquaintance with them justifies.

One way in which evidence of gene arrangement within chromosomes may be secured has been extensively used for many years. When in an experimental colony of animals of known stock there appears an individual which differs in some definite physical respect from its parents, it is called *a mutant*. If physical peculiarities depend on genes within chromosomes, theoretically it should be possible by critical study of the dividing germ cells of the mutant to find atypical chromosomes. There should be some change from the usual shape due to a gene not usually present, or a minute gap where a gene usually present is missing. In some of the experimental forms, such as the fruit fly, Drosophila, which has been intensively studied from the genetic standpoint, this has proved to be the case. Even more striking than the evidence obtained from spontaneous mutants has been that obtained more recently by irradiation of the germ cells. If during their maturation germ cells are subjected to x-rays of an intensity not quite sufficient to kill them, some of the genes are destroyed or displaced. When animals so treated are bred, some of their offspring show definite changes in bodily structure. In other words mutations have been artificially induced which may be correlated with equally definite changes in the gene composition of particular chromosomes. The work of thus tying up specific hereditary traits with definite chromosomes, though as yet in its infancy, has already yielded most interesting results. We are beginning to see more clearly the manner in which both the racial heritage and the particular traits of an individual are passed on. The links in the endless chain of heredity are the gene-bearing chromosomes -a definite number of pairs maintained constant in all the cells of an individual by mitosis; the same definite number of pairs in the cells of each succeeding generation reestablished when each gamete brings in one member of the new pair.

The final two divisions in spermatogenesis and oögenesis are not typical mitotic divisions. To distinguish them they are called *meiotic divisions, or* less technically, *maturation divisions.* In these divisions the chromosomes of the gametes are reduced to half the species number. The process of thus halving the species number of chromosomes is called *meiosis.* Cytologists have worked out the mechanism of these maturation divisions with great care in many forms. In some forms chromosome reduction takes place in the first, in others in the second maturation division. Which of the two divisions is a reduction division in any particular case is of minor significance. The reduction itself is the thing of interest. Stripped of detail and without reference to many peculiar modifications encountered in different animals, a *reduction division is* a cell division in which the chromosomes are not split in the metaphase. Instead the chromosomes are redistributed, half of them moving bodily to one daughter cell and half to the other.

In this halving process the way is paved for separating the chromosomal pairs by a special mechanism termed *synapsis* which occurs in the prophase of a maturation division but not in an ordinary mitosis. If we look at the spindle of an ordinary mitosis we can recognize the members of the chromosomal pairs by their size and shape, but the members of the pairs do not lie next to one another. When, during the prophase, the spireme thread was broken up into chromosomes these chromosomes aggregated at once in haphazard arrangement at the equator of the spindle. In the prolonged prophase of the first maturation division the members of the chromosomal pairs come to lie close to each other and so remain for some time. This pairing off of the chromosomes in the prophase is what is meant by the term synapsis. These synaptic pairs of chromosomes, still in intimate association, finally move to the equator of the spindle. Whereas in an ordinary mitosis splitting of the chromosomes would occur when they had arrived at the equatorial position, in a reduction division the two members of the synaptic pair are separated from each other, one going bodily to either pole of the spindle. The resulting cells thus receive half the species number of chromosomes, and this half-complement *(haploid number) is* made up of one member of each of the pairs characteristically present in the species *(diploid number).* If the first of the maturation divisions has been a reduction division, as just described, then the second maturation division will be an *equation division.* In it the chromosomes split longitudinally yielding, as the designation equation division implies, two daughter cells, each with the same haploid number of chromosomes exhibited at the beginning of the division.

In many forms which have been carefully studied the synaptic pairs of chromosomes, as the first maturation division is approached, exhibit a peculiar quadripartite appearance. This has led to their being designated as a *tetrad.* This condition is due merely to the fact that internal splitting of the chromosomes has become apparent before their parts are widely separated by movement toward the poles of the spindle. Thus a tetrad is nothing but a synaptic pair of chromosomes in which each member of the pair is showing a sort of precocious internal separation. This same type of early division can be observed in ordinary mitoses but it is not usually as conspicuous as in the first maturation division.

If we recall, now, what we learned of the gene composition of chromosomes, it will be evident that the cells formed in the maturation divisions contain different hereditary potentialities because they contain different chromosomes, not halves of the same chromosomes as results in an ordinary mitosis. What hereditary possibilities are discarded into the polar bodies thrown off from the femal gamete and what are retained in the mature ovum is a matter of chance distribution. What potentialities find their way into the particular sperm which alone, out of millions of its fellows, fertilizes the ovum is likewise fortuitous.

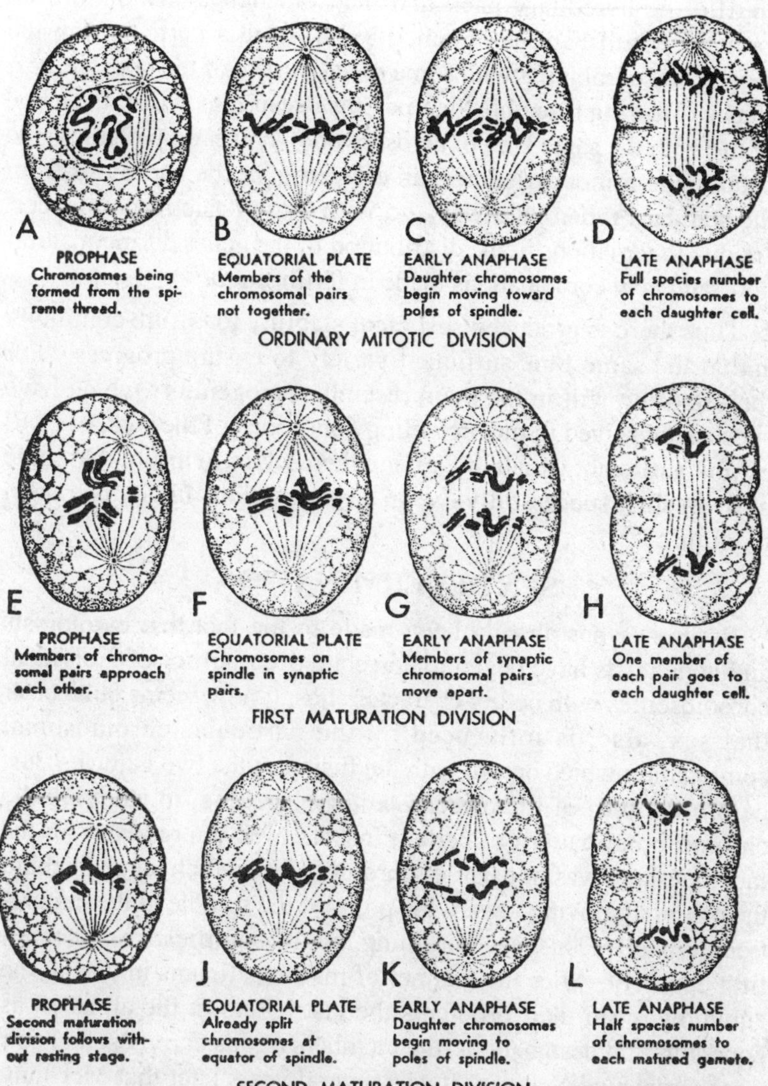

Figure 2.7 : Diagrams showing schematically the difference between a reduction division in the maturation and an ordinary mitotic division. For the sake of simplicity of representation the species number of chromosomes is assumed to be eight.

"Thus in the game of life, the maturation processes virtually shuffle the hereditary pack and deal out half a 'hand' to each gamete. A full hand is obtained by drawing a partner from the 'board,' by combining with some other gamete of the opposite sex. Hence offspring resemble their parents because they play the game of life with the same kind of cards, but not, however, with the same hands. The minor differences in offspring, or the variations from the standard type that always go with these basic resemblances, are due to variations in the distribution of the genes during maturation" and new combinations made in fertilization.

Thus there is produced sufficient stability to ensure continuity and at the same time sufficient variety to ensure progress. "For the offspring will in the main resemble. progenitors which have successfully lived in the prevailing conditions of the past, but will exhibit sufficient variability among themselves to insure that some of them shall successfully live in any conditions likely to arise in the future."

SEX DETERMINATION

Reference has already been made to the fact that cytologists and geneticists have begun to correlate peculiarities of individual chromosomes with bodily characteristics. It now seems quite clear that sex, also, is influenced by the particular chromosomal combination which occurs with the fusion of the two gametes. Just before the turn of the present century Henking, in studying the chromosomal pattern of certain insects, was impressed with the fact that there was one pair of chromosomes which lagged behind the others in moving toward the poles of the spindle in the maturation divisions. This was intriguing, but its significance was not it first apparent. After the manner of mathematicians in giving the "unknowns" in their problems the last letters in the alphabet as noncommittal designations, the members of this chromosomal pair were christened X and Y. It was several years later that McClung and Wilson came to the conclusion that this pair of chromosomes was involved in sex determination. In line with this pioneering work, practically all the animals which have been critically studied show a slight but strikingly consistent difference in the chromosome picture exhibited by the cells in the bodies and in the germ line of the two sexes. In one sex, the cells have all their pairs of chromosomes symmetrically mated. In the opposite sex one of the pairs of chromosomes is likely to be peculiar in that its members are quite different in size and shape instead of being like each other as is the general rule. Subsequent experimental work has supported the conclusions of McClung and Wilson and given us considerable evidence indicating that the reason the members of the X-Y pair are unlike is because something that makes for femaleness is carried in the X chromosome and absent in the Y.

We have seen that in the reduction division of the gametes the members of the chromosomal pairs are separated. If the X chromosome does carry whatever makes for femaleness and its Y mate lacks this or carries an antagonistic male determining factor, it is evident that when in a reduction division the members of this chromosomal couple go to opposite poles of the spindle the resulting cells will carry opposite sex tendencies.

Assuming (as actually is the case in most of the forms so far studied') that it is the female which has all its chromosomes symmetrically paired and the male that has an unbalanced pair, the chromosomal combinations taking place in fertilization can readily be depicted graphically. When the female gametes mature, all of them will appear alike as far as chromosomes are concerned, because in maturation one member of each balanced pair goes to each cell. The reduction division in spermatogenesis, however, separates the unbalanced X-Y pair and consequently half the gametes receive an X chromosome and the other half a Y chromosome. The chances of an X-carrying gamete uniting with an ovum and giving rise to a female are the same as those of a Y-bearing chromosome fertilizing the ovum to produce a male. Thus this theory accounts at the same time for the approximate equality of the numbers of males and females produced, and for the observed differences in the chromosome picture presented by the two sexes.

Recent experimental evidence seems to indicate that there are other factors involved in sex determination, and that sex may not

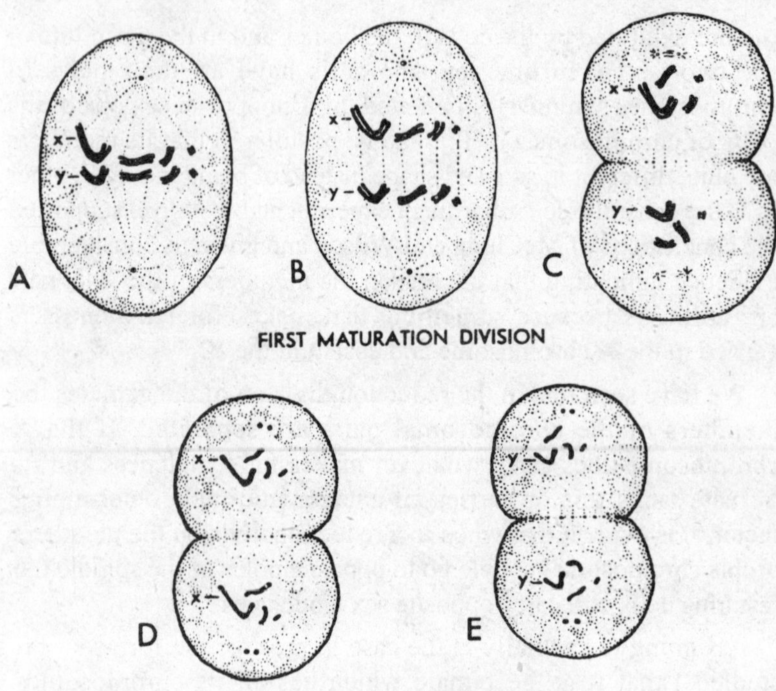

Figure 2.8 : Schematic diagrams indicating the manner in which the chromosomes believed to be involved in sex determination are distributed in the maturation divisions.

be as irrevocably fixed at the time of fertilization as was at first believed. There seems little doubt that the chromosomal combination is effective in starting the trend of development toward one sex or the other, but apparently this trend may be inhibited or modified by massive doses of male or female sex hormones during the period when the gonads are being differentiated. Moreover, from his work on intersexes, Bridges believes that there are genes in other chromosomes than the X-Y pair which influence the development of maleness or femaleness. When these factors are in balance, the XY or XX combination is sufficient to throw the course of development to one sex or the other Unbalance of the ancillary factors from other chromosomes under certain circumstances may operate against this determining influence of the s called "sex chromosomes" with the resultant production of

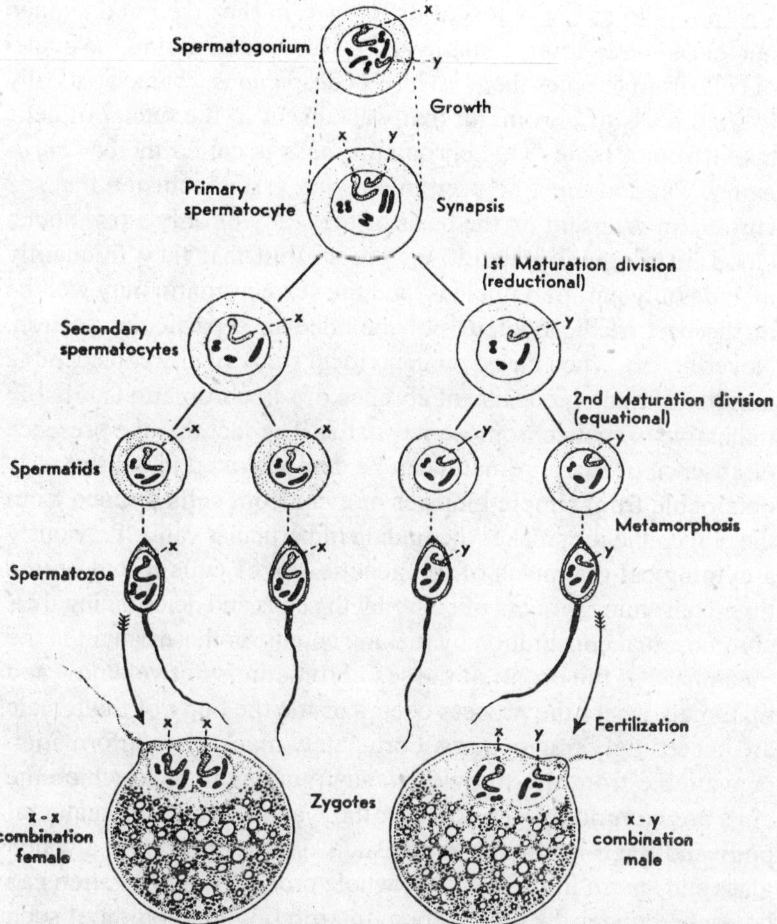

Figure 2.9 : Schematic diagram showing the separation of the members of the sex chromosome pair in maturation and their recombinations in fertilization. It is assumed that the species number of chromosomes is eight and that it is the male which produces gametes of different potentialities with regard to sex determination. The sex chromosomes are stippled; other chromosomes are drawn in solid black.

sex intergrades. Whatever future work may show the accessory or modifying factors to be, it seems quite clearly established that the X-Y chromosomal pair plays an important, probably the major, role in the determination of sex.

In 1949 Barr and Bertram first presented evidence that it +/

Vas possible to see a sexual difference in the fixed and stained nuclei of nondividing somatic cells. They found that in the nuclei of cells from females there usually a conspicuous, characteristically located mass of chromatin that was absent in the nuclei of cells taken from a male. This chromatin mass is called the *sex chromatin.* Caution must be used in arriving at a conclusion that sex chromatin is absent on the basis of the study of only a few nuclei Good histological preparations are so thin that they frequently include only part of a nucleus, and the sex chromatin may well be in the part of the nucleus not included in a particular section. Nevertheless, when based on the critical study of any considerable number of nuclei, consistent absence of sex chromatin is reliably indicative of male chromosomal status. The fact that the presence or absence of sex chromatin can be demonstrated in cells readily obtainable from simple biopsies or even from cells scraped from the oral epithelium makes the finding of particular value. Previously a cytological diagnosis of the genetic sex of cells rested on the time-consuming process of culturing the cells and determining their chromosomal constitution by making counts on those caugnt in me metaphase of the origin of the sex chromatin is not yet know and mammals where the process occurs inside the body of the female are exceedingly fragmentary. Fortunately much more information is available from the study of water-living animals in which the eggs are commonly deposited in the water outside the maternal body and there fertilized. When such eggs are placed in a watch glass and sperm introduced, the whole process of fertilization can be studied under the microscope. Interpreting in the light of such observations we can piece together a generalized account of the process of fertilization which in all probability is applicable without essential modification to birds or mammals.

When liberated from the ovary, an ovum at once begins to undergo certain changes which can be characterized as aging or deterioration. Among other things there is a tendency for its protoplasm to become progressively more coarse. With this loss of the originally finely dispersed phase of its colloidal material the ovum loses vigor. These changes progress rapidly to a point where the ovum, although technically still alive, can no longer be fertilized.

If, however, the ovum is fertilized reasonably promptly after its liberation, these deteriorative changes are checked and the protoplasm increases its activity in a way that is often described as "being rejuvenated." The nature of these changes is not as yet fully understood, but they involve increase in permeability and increase in oxidation rate. There is also an increase in the amount of ammonia excreted. This indicates that there has been an increase in purine metabolism, which is important in the synthesis of nuclear materials. These changes symbolize, of course, the beginning of the period of tremendously rapid growth which is destined to end in the production of a new individual.

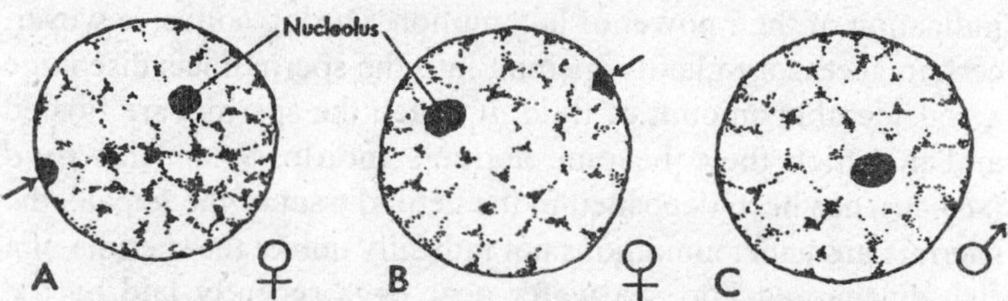

Figure 2.10 : Drawings of the nuclei of human epidermal cells to show the sex chromatin. (A) and (B) Nuclei from a female, the sex chromatin indicated by arrows. (C) Nucleus from a male, showing the absence of sex chromatin. Note that the nucleolus, although eccentric in position, does not lie against the nuclear membrane.

Interestingly enough, fertilization in the usual manner by the male sex cell is not the only way an ovum may have its deterioration checked and be started on its growth phase. A variety of other stimuli may be substituted for the male sex cell. Changes in the ionic concentration of the sea water is effective in initiating development in the eggs of some marine invertebrates. Insect eggs have been started developing by stroking them with a camel's hair brush. Pricking with a needle has been effective in inducing frog's eggs to begin to grow. The experimental initiation of development in an egg by means other than its union with the male sex cell is known as artificial parthenogenesis. The fact that artificial parthenogenesis is possible emphasizes the fact that all the necessary morphogenetic factors for development are present in the ovum and that the process of fertilization is essentially, as Barth puts it, a sort of "release mechanism." In other words something from the sperm cell, or in artificial parthenogenesis the chemical or mechanical factor employed, releases potential energy already present in the egg. It should not be overlooked, however, that this release mechanism is far from the whole story. The sperm brings with it its quota of hereditary potentialities, and the mixture of germinal material from two parents produces a vigor of growth not manifested in eggs starting their development by artificial parthenogenesis.

Turning now from the ovum to the male sex cells, it is interesting that in the testes and sperm ducts the spermatozoa show but little indication of their power of locomotion. During coitus, however, certain accessory glands opening into the sperm ducts discharge a considerable amount of fluid in which the spermia are floated and in which they' become actively motile. When this *fluid* (*semen*), has been deposited in the genital tract of the female, the spermia are under conditions not radically unlike the spermia of a fish discharged into sea water near eggs recently laid by the female. The bird ovum, recently liberated from the ovary, is awaiting the spermia in the upper part of the oviduct. Here the ovum becomes surrounded by swarms of spermatozoa which have made their way thither by their own active locomotion through the lower part of the genital tract.

The process to this point is known as *insemination.* Still other preliminary steps are to be passed through before the nuclei of the male and female gametes unite to complete the process of fertilization. The spermia once having reached the neighborhood of the ovum tend to remain there held by chemical interactions which are not as yet fully understood.

Apparently, at least in the case of some of the most critically studied of the marine invertebrates, both the ova and the sperm produce substances having definite effects on each other. These substances are in some ways similar to hormones, and because they are produced by gametes, they have been called *gamones.* The active substances produced by ova are known as *gynogamones* and the comparable substances resident in sperm are called *androgamones*. Work in this field is relatively new, and the conclusions from it must still be regarded as tentative. On the basis of current interpretations if appears as if there were two gynogamones. Gynoganome I is believed to activate the sperm cells to vigorous swimming movements. Gynogamone II makes the surface of the spermatozoa heads sticky, and may thus be a factor in adherence of a sperm cell to the surface of the egg cell for sufficient time to initiate penetration. The fact that it also causes sperm cells to stick together, or agglutinate, seems incidental.

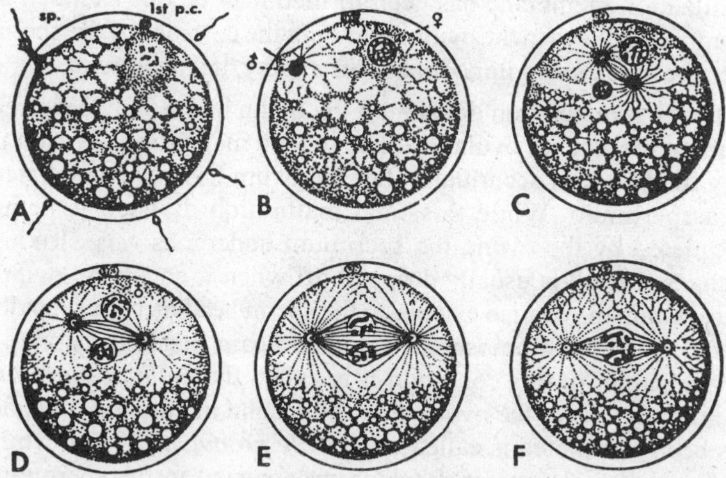

Figure 2.11 : Diagrams schematicallyu illustrating the process of fertilization and the formation of the first cleavage spindle. The species number of chromosome is assumed to be eight. Abbreviations: sp., spermium; p.c., polar cell (polar body). The male and female symbols designate the male and female pronuclei. respectively.

There seem to be, also, two androgamones. Androgamone I is believed to inhibit active sperm movement. Spermatozoa contain a minimal amount of energy-producing materials so that their potential kinetic energy is strictly limited. Inactivity until the critical time arrives is, therefore, important and it appears as if androgamone I enforces such inactivity until its action is nullified by the activating effects, first of the male accessory gland secretions and then by gynogamone I. Androgamone II appears to be responsible for the dissolving of the outer egg membranes, thereby facilitating the penetration of sperm cells.

When a sperm cell has penetrated the outer investment of the egg, a minute projection of ovarian cytoplasm, called the fertilization cone, rises up to meet it and draws it into the ovum. Not infrequently, especially in birds, several spermatozoa penetrate the ovum, but since only one of these taxes part in fertilization we can neglect the other sperm has penetrated the ovum the peripheral cytoplasm produces a clear viscid substance which adheres to the inner face of the vitelline membrane, thickening it into the so-called *fertilization membrane.* The fact that no spermia penetrate the ovum after the fertilization membrane has been formed may be due as much to chemical changes in the ovum following the entrance of the sperm as to the mechanical impediment offered by the membrane.

The first maturation division of the ovum has usually occurred at about the time of ovulation. The second maturation division is very likely not to occur until after the ovum has been penetrated by a spermium. While this final maturation division is being completed by the ovum, the spermium undergoes *very* striking changes. Its tail is usually dropped off when it enters the ovum. Once within the ovarian cytoplasm the sperm head increases rapidly in size and its chromosomal contents again become distinctly recognizable. In this condition it is called the *male pronucleus,* and the nucleus of the ovum after the second maturation division has been completed is called the *female pronucleus.* Meanwhile the centrosomal apparatus, which was carried in the spermium where its tail was attached to its head, becomes much more conspicuous. As it approaches the female pronucleus, the male pronucleus with its associated centrosome rotates so that the centrosome moves ahead of the chromatin material. By the time the two pronuclei are close to each other this centrosome has divided and formed a mitotic spindle on which the chromosomes brought in by the sperm and those in the ovum both aggregate. Fertilization can be regarded as complete when the maternal and paternal chromosomes are thus grouped together ready to be split in the impending first cell division in the life of the new individual.

FORMATION OF THE ACCESSORY COVERINGS OF THE OVUM

Fertilization, as we have seen, normally takes place just as the ovum is entering the oviduct. The accessory coverings, as the albumen, shell membrane, and shell are called, are secreted about the ovum during its subsequent passage toward the cloaca. In the part of the oviduct adjacent to the ovary a mass of stringy *albumen is* produced. This adheres closely to the vitelline membrane and projects beyond it in two masses extending in either direction along the oviduct. Due to the spirally arranged folds in the walls of the oviduct, the egg as it moves toward the cloaca is rotated. This rotation twists the adherent albumen into the form of spiral strands projecting at either end of the yolk, known as the *chalazae*.

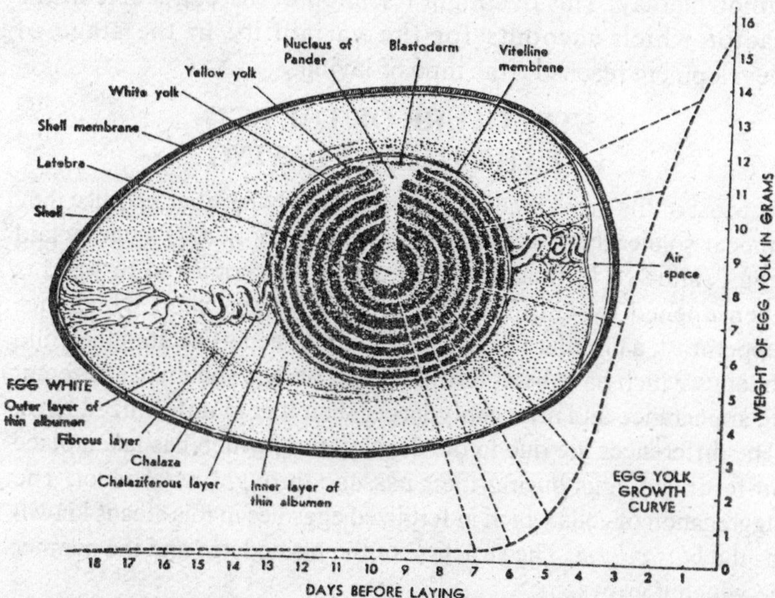

Figure 2.12 : Diagram showing the structure of the hen's egg at the time of laying. The graph indicates the rate of growth of the egg during the 18 days preceding its laying. The lines leading from the various layers of the yolk to the growth curve emphasize the time at which these layers were formed.

Additional albumen, which has been secreted abundantly in advance of the ovum by the glandular lining of the oviduct, is caught in the chalazae and during the further descent of the ovum is wrapped about it in concentric layers. These lamellae of albumen may be easily demonstrated in an egg which has had the albumen coagulated by boiling. The albumen-secreting region of the oviduct constitutes about onehalf of its entire length.

The *shell membranes* which consist of sheets of matted organic fibers are added farther along in the oviduct. The *shell is* secreted as the egg is passing through the shell-gland portion of the oviduct. The entire passage of the ovum from the time of its discharge from the ovary to the time when it is ready for laying has been estimated to occupy about 22 hours. If the completely formed egg reaches the cloacal end of the oviduct during the middle of the day it is usually laid at once, otherwise it is likely to be retained nut: the following day. This overnight retention of the egg is one of the factor which accounts for the variability in the stage of development reached a the time of laying.

STRUCTURE OF THE EGG AT THE TIME OF LAYING

Most of the gross relationships are already familiar because they appear so clearly in eggs which have been boiled. If a newly laid egg is allowed to float free in water until it comes to rest and is then opened by cutting away the part of the shell which lies uppermost, a circular whitish area will be seen to lie atop the yolk. In eggs which have been fertilized, this area is somewhat different in appearance and noticeably larger than it is in unfertilized eggs. The differences are due to the development which has taken place in fertilized eggs during their passage through the oviduct. The aggregation of cells which in fertilized eggs lies in this area is known as the *blastoderm.* The structure of the blastoderm and the manner in which it grows.

Close examination of the yolk will show that it is not uniform throughout either in color or in texture. Two kinds of yolk can be differentiated, *white yolk* and *yellow yolk.* Aside from the difference in color visible to the unaided eye, microscopical examination will show that there are differences in the granules and globules of the two types of yolk, those in the white yolk being in general smaller and less uniform in appearance. The principal accumulation of white yolk lies in a central flask-shaped area, the *latebra,* which extends toward the blastoderm and flares out under it into a mass known as the *nucleus of Pander.* In addition to the latebra and the nucleus of Pander there are thin concentric layers of white yolk between which lie much thicker layers of yellow yolk. The concentric layers of white and yellow yolk are said to indicate the daily accumulation of deutoplasm during the final stages in the formation of the egg. The outermost yolk immediately under the vitelline membrane is always of the white variety.

The albumen, except for the chalazae, is nearly homogeneous in appearance, but near the yolk it is somewhat more dense than it is peripherally. The chalazae serve to suspend the yolk in the albumen.

The two layers of shell membrane lie in contact everywhere except at the large end of the egg where the inner and outer membranes are separated to form an *air space.* This space is stated (Kaupp) to appear only after the egg has been laid and cooled from the body temperature of the hen (about 106° F) to ordinary temperatures. In eggs which have been kept for any length of time, the air space increases in size due to evaporation of part of the water content of the egg. This fact is taken advantage of in the familiar method of testing the freshness of eggs by "floating them."

The egg shell is composed largely of calcareous salts. These salts are derived from the food of the mother, and if lime-containing substances are not furnished in her diet, the shell is defectively formed or even altogether wanting. The shell is porous allowing the embryo to carry on exchange of gases with the outside air by means of specialized vascular membranes arising in connection with the embryo but lying outside it, directly beneath the shell.

INCUBATION

When an egg has been laid, development ceases unless the temperature of the egg is kept nearly up to the body temperature of the ,mother. Moderate cooling of the egg does not, however, result in the death of the embryo. It may resume its development if it is brooded by the hen or artificially incubated even after the egg has been kept for several days at ordinary temperatures.

The normal incubation temperature is that at which the egg is maintained by the body heat from the brood-hen. This is somewhat below the blood heat of the hen (106° F). When an egg is allowed to remain undisturbed, the yolk rotates so that the developing embryo lies uppermost. Its position is then such that it gets the full benefit of the warmth of the mother.

In incubating eggs artificially the incubators are usually regulated for a heat of 99 to 100° F (36 to 38° C). At this temperature the chick is ready for hatching on the 21st day. Development will go on at considerably lower temperatures, but its rate is retarded in proportion to the lowering of the temperature. Below about 21° C development ceases altogether.

If eggs which have been cooled after laying are to be incubated for the purpose of securing embryos of a particular stage of development, 3 or 4 hours are ordinarily allowed for the egg to become warmed to the point at which development begins again. For example if an embryo of "24-hours' incubation age" is desired, an egg that has been subjected to cooling should be allowed to remain in the incubator about 27 hours. Even with allowance made for the warming of the egg and with exact regulation of the temperature of the incubator, the stage of development attained in a given incubation time will vary widely in different eggs. The factor of individual variability, which must always be reckoned with in developmental processes, undoubtedly accounts for some of the variation. The different lengths of time occupied by eggs in traversing the oviduct, the overnight retention of eggs not ready for laying until toward sundown, and especially the varying time different eggs have been brooded before being removed from the nest account for further variations. The designation of the age of chicks in hours of incubation is, therefore, not exact, but merely a convenient approximation of the average condition reached in that incubation time.

3

Incubation Requirements

When the egg is laid development has already begun. It is then suspended until the egg is warmed again sufficiently. The time between this renewed warming and emergence of the chick (the incubation period) varies between species, and is roughly in proportion to egg size. In small song birds the incubation period can be as short as 10-14 days, in the domestic fowl it is 20-21 days and it extends to 42 days in a large sea bird such as the gannet or up to 50 days in the Mallee fowl and the brush turkey while in the royal albatross it lasts about 80 days.

In all species the rate of development and the level of hatchability vary with the conditions of incubation, one of the more important of these being temperature. For obvious economic reasons the requirements of the developing embryo - the optimum conditions of incubation - have been assessed in the greatest detail in eggs of the domestic fowl and in artificial incubators. Work in this field (the consideration of incubator conditions associated with good hatchability) has been summarized by Landauer (1987) and by Lundy (1989). Under good incubation conditions and after being held under the correct storage conditions (Proudfoot, 1989) eggs from good stock can be expected to give a hatchability of well over 90% of fertile eggs set. In this chapter the conditions within artificial incubators which are consistent with a high level of hatchability will be considered first for the fowl, and then the effects of these conditions on other domesticated species. Incubation procedures of a number of species of wild birds will then be discussed.

REQUIREMENTS OF THE EMBRYO OF THE DOMESTIC FOWL

Heat

In a forced draught incubator an optimum incubation temperature has been found to be 37·8°C. There is, however, evidence that hatchability is unaffected if the incubator temperature is dropped by 1½°C after 16 days of incubation; and at this time it is known that embryo temperature has begun to rise above that of the incubator. Research of a number of Russian workers, summarized by Lundy (1989), claims that fluctuations in incubator temperature provide an added increase in hatchability.

Humidity

Incubator humidity determines the rate of water loss from the egg. Maximum hatchability can be obtained with a relative humidity between 40% and 60%. However, a reciprocal relationship has been found between temperature and humidity, such that if the humidity level is raised the incubation temperature can be lowered and vice versa.

The Gaseous Environment

Correct ventilation of the incubator is as important as its temperature and humidity. With regard to the oxygen content of the air there may be an optimum concentration for maximum hatchability at 21 %. With regard to carbon dioxide the picture is still unclear. It is known, however, that a level of carbon dioxide above 1 % results in a decrease in both hatchability and growth rate, and also that the sensitivity of the embryo to carbon dioxide decreases with age. During late embryonic development a rise in the Pa_{CO_2} of the blood has been found by Visschedijk (1998c) to be necessary for normal development.

Egg Position and Change of Position

Eggs must be turned periodically, and egg-turning is particularly important between the fourth and seventh days of incubation. Lack of turning can result in premature adhesions between the extra-embryonic membranes and distortions in subsequent development. The amount of turning also appears to be important, a minimum being about three times a day while more than 24 times a day is unnecessary; adequate turning reduces mortality especially between days 5 to 16 and 19 to 21. When turned, eggs should not be rotated always in the same direction as this can result in rupture of the yolk sac, disruption of the chorion, allantois and shell membranes, twisting of the chalazae and rupture of the blood vessels. In incubators with mechanical turning, the egg is kept upright and the tray is moved alternately from side to side through an angle of about 65°.

It has been found important also to incubate eggs either horizontally, or with the large end upwards. With this limitation turning of the eggs in a number of different planes of rotation has been shown to have a beneficial effect on hatchability. When eggs are incubated with the small end uppermost there is an increase in embryonic malpositioning (head in the small end of the egg) which leads to a reduction in hatchability.

INCUBATION REQUIREMENTS OF OTHER DOMESTIC SPECIES

For reasons which will become clearer when incubation under natural conditions is considered below, hatching success, which is high when eggs of the fowl are incubated artificially, may fall off when the same conditions are provided for other species. Where other domesticated species are concerned the difficulties appear to arise from levels of humidity more than any other factor. Although the differences here are not fully understood, practices recommended for the artificial incubation of duck, goose and turkey eggs usually include a level of humidity higher than that required by the fowl. It is, for instance, frequently recommended that eggs of the duck and goose should be sprinkled with water at intervals during incubation. Insko (1991) recommends a higher level of humidity for turkey as compared with chicken eggs for the first part of incubation (61-63% as compared with 55-61%), and 65% for the first 24 days, rising to 70-75% for the last few days has also been recommended (Ministry of Agriculture, Fisheries and Food, 1960). With regard to temperature Romanoff (1935) found hatchability to be best in the turkey between 36° and 38°C, and Martin & Insko (1985) obtained their highest hatchability (in still air incubators with the bulb of the thermometer level with the top of the eggs) with average temperatures of 38·1°, 38·6°, 39·2° and 39·4°C for the first, second, third and fourth weeks of incubation respectively.

The level of humidity recommended for ducks is higher than that for the turkey. For the domestic duck 70% is suggested for the first 24 days, then 60% until pipping, and then a return to the original level. For the runner duck Romanoff (1993) found the optimum temperature to be 37·4°C, with a drop of 0·3°C at the time of hatching, and the optimum relative humidity to be 70% or higher until day 24 and 60% or lower after that. An even higher level of humidity has been found the most effective for the game farm mallard duck; Prince *et al.* (1999) obtained their best hatch with a temperature of 37·5°C and -a humidity level of 70-80%. Kaltofen (1991) also raised the incubator humidity to 80% at the time of pipping, for the Pekin duck.

In geese the recommended temperature is 37·2°C with a relative humidity of 70%. Romanoff (1984) considered whether the eggs of game birds (in this case the ring-necked pheasant and the bobwhite quail) could be incubated together successfully. Experimental results showed this to be inadvisable as, in a still air incubator with the temperature recorded at the level of the upper side of a hen's egg, the optimum temperatures for the two species were different. Hatchability in the pheasant was highest at an incubator temperature of 38·9°C for the first week, 38·3°C for the second and 37·8°C for the third, while it was 38·3°C throughout incubation for the quail, with a permissible small rise towards the time of hatching. The optimum humidity for the pheasant fell from 75% at the beginning of incubation to 65% at the end, while in the quail it rose from 65% at the beginning to 75% at the end. For the quail the permissible range was found to be narrower than for the pheasant, in respect to both temperature and humidity.

INCUBATION IN WILD BIRDS

Incubation requirements of wild birds are very similar to those of the fowl. But, whereas work on incubation in the fowl arises from attempts (extending over several thousand years) to incubate eggs artificially, what we know about incubation in wild species arises largely from research on the breeding biology of specific species, in which the mode of adaptation within their ecological niches is usually considered. An attempt will be made to bring information on wild species into the focus provided by research on the fowl. From this it appears that incubators eliminate two types of selection pressure which affect wild species. Thus, incubation in the wild is adapted partially as a result of climatic pressures (as from extreme cold or heat), and also as a result of predation, which can threaten the parent, and the eggs. Bearing these additional hazards in mind, we shall be concerned briefly with the problem of how the physical requirements of incubation are met under natural conditions in a few selected species.

Incubation Sites

Nest- or incubation-sites are almost infinitely variable between species, descriptions of these are readily available in the ornithological literature and a review of incubation practices is outside the scope of the present volume. Therefore incubation sites will be described for a few species where there is much information or which raise specific issues to be dealt with below.

Guillemots have no nest. They breed in densely packed colonies on the ledges of steep cliff faces, and the single egg is incubated on the bare rock. It is shaped in such a way as not to roll except in a half circle (the ledges may be very narrow). The egg is tended continuously, by one parent or the other. The parents incubate in rows, shoulder to shoulder. In this way they present a barrier against predation which is noticeably weakened if, for some reason, one incubating pair is lost.

Gulls are also colonial nesters, in this case on the ground and each pair defends a territory around its nest (for the herring gull see Tinbergen, 1993; Drent, 1990, and for the blackheaded gull, Beer, 1991). Tinbergen (1993) describes the herring gull nest as a well-rounded and well-lined nest cup, partly scraped out of the ground and partly built up of straws and moss which are arranged in it, or round the edge. Again, incubation is undertaken by both sexes, they incubate continuously (about 95% of the time; Drent, 1990), and the eggs are normally left only when the birds fly up in response to intruders. Under normal conditions three eggs are laid.

The *yellow-wattled lapwing* nests in the dry season, in tropical conditions. It is a solitary species confined to Pakistan, India and Ceylon. The breeding behaviour has been described by Jayakar & Spurway (1985a, b). The pairs watched by them nested in the neighbourhood of water. The nest is a small concavity surrounded by a small circular bank of gravel and twigs. Four eggs are laid in each nest. One or other of the parents remains on the nest during the night. During the daytime the eggs are left for long periods apparently unattended, although one or both adults invariably appear if the nest is approached by intruders.

The breeding biology of the *greater crested grebe* has been described by Simmons (1995). In this species the nest-site appears to be dictated by the threat of predation. Grebes are exclusively aquatic birds and very clumsy on dry land. They are expert divers and the nest is sited in or near the water, and may be partially floating on it. When approaching, the parent swims right up to it. It is built of sodden rotting weed. During the laying period, or when approached by intruders, the incubating bird swiftly covers the eggs with this material before sliding into the water. The nests are frequently awash, and when the water splashes over the rim they are built up with more weed. The egg surface consists of a white chalky substance which covers the blue shell.

Like many other small song birds, the *wren* builds nests with a quality noticeably lacking in those described above, that is, insulation. Work on nest-building in the wren has been reviewed by Armstrong (1995). The structure, which may be built into a cavity, or into surrounding twigs on a stump, is roofed and has an entrance hole in the side. It is built by the male, often using materials which blend into the background sufficiently to make it hard to locate, and is lined with 'feathers and other downy material' by the female. The eggs are incubated exclusively by the female, who occupies the nest intermittently in the daytime. The insulating properties of nests have been considered by Drent (1992), together with the energy cost which the process of incubation imposes on the parent and the possible part played by the nest in reducing this.

In *megapodes*, such as the Mallee fowl, and the brush turkey the use of insulation is carried to its extreme. In both species the male constructs a mound of mixed earth and leaves which then heats up. In the brush turkey it is built at a rainy period, and is about 4 m wide and 1 m high. Before laying each egg the female excavates a hole in the mound, deposits the egg in an upright position and covers it over with the nest material. A clutch can consist of about 24 eggs laid at intervals over several weeks.

Warming of the Eggs

Data collected by Drent (1992, 1993) show that incubation temperatures do not vary between species by much more than 4°C. He points out that this similarity is an expression of the similarity in body temperature in adults of different avian species.

In almost all species the eggs are brought up to the incubation temperature by contact with the parent; in most species one or both parents develops a brood patch, or patches. Brood patches are vascular defeathered, and later, oedematous areas which appear on the bird's ventral surface at the time of egg laying. Hormonal changes underlying this development are discussed by Hinde (1997) and by Jones (1991). There are many descriptions of ways in which incubating parents bring their brood patches into the right relationship with the egg or eggs, see for instance Simmons (1995) for the grebe, Drent (1990) for the herring gull and Beer (1991) for the blackheaded gull. Certain species, for example cormorants, do not develop a brood patch, and in some, such as the gannet and the booby incubation is carried out by enclosing the single egg between the webbed feet. According to Howell & Bartholomew (1992) the mean internal temperature of the egg of the red-footed booby, similarly incubated, is 36°C and the foot temperature 35·8°C. In megapodes the heat is provided by the process of fermentation, supplemented, in the Mallee fowl by the heat of the sun. The mound of the brush turkey is tended throughout the incubation period by the male parent, which periodically excavates 'control' shafts : these are holes scratched in the mound, about 20-30 cm deep, in which the male sticks its head for a few seconds, before refilling them. The behaviour of the Mallee fowl is similar, except that in the arid scrub lands where it lives, leaves are scarce and the mound is constructed largely of sand. For this reason fermentation is less and Frith (1982) reports that three types of work are required from the male to keep the temperature constant. Early in the season, when fermentation is rapid the mound temperature rises rapidly. The male opens the mound every morning, to within a few inches of the eggs, fills in the hole and leaves. This results in a drop in temperature. Later fermentation is slower, fewer holes are excavated, and the eggs are covered with a foot or two of soil. Later still, the rate of fermentation decreases again, but the sun becomes hotter. The bird then works to keep the mound cool by piling soil over it, sometimes to a height of about 11 m. Sometimes the mounds are completely dug out in the early morning, the sand spread out to cool, then piled up again. According to Baltin (1969) the mean temperature in a brush turkey mound is between 33·3°C and 33·9°C; beginning at about 36·7°C it falls through the five months of incubation to about 28·8°C.

In both the last two families (Megapodiiadae and Sulidae) the egg temperature is more likely to resemble that of an egg in a forced draught incubator, in the sense that heat is applied rather evenly over its surface. In most species the reverse is the case; a normal nest resembles more a still air incubator, where the upper surface of the egg is maintained at the higher temperature. For example, Baldwin & Kendeigh (1982) recorded temperatures at different places in the wren's nest. They found that the highest temperature is in the area where the egg makes contact with the brood patch. The outside of the clutch is held at an intermediate temperature, and the nest bottom is considerably cooler. This situation is largely reversed when the incubating parent leaves the nest.

Egg-shifting behaviour resulting in reorganization of the clutch, may also counteract temperature gradients within the nest. There is more evidence that incubation is not an all-or-nothing affair: the parents' incubation behaviour varies with egg temperature. Franks (1997) has observed the responses of incubating ringed turtle doves to artificial eggs where the temperature could be changed. When the egg temperature was high gular flutter occurred, and elevation of the feathers and shivering was observed when egg temperatures were lowered. Gular flutter is a mechanical aid to evaporative cooling. Baerends *et al.* (1990) and Drent *et al.* (1990) have demonstrated that egg temperature is one of the feed-back stimuli regulating incubation behaviour. They have shown that the herring gull's response to changes in egg temperature is aimed at maintaining the parent's body temperature. In this way egg temperature is regulated while allowing for almost continuous coverage of the eggs.

The Problem of Cooling

The problem faced by incubating birds in hot climates is a very specific one which has been touched on already at certain stages in the megapodes. It is discussed by Drent (1992); in his view the response of the parent to extreme heat and exposure to the sun is to remain at the nest; the problem of keeping the eggs cool is thus shifted to the problem of keeping the parent cool. Drent points out that different ways of doing this have evolved in different species and include the use of poorly feathered body regions which can act as dissipators of heat, panting or gular flutter, the periodic wetting of the abdominal feathers and the orientation of the sitting bird away from the sun. It appears that at extremes of air temperature the eggs cannot be protected by shading alone, but may have to be cooled by contact with the brood patches. For example, in the double-banded courser Maclean (1997) found that the single egg could be left unattended at air temperatures between 20° and 30°C, was frequently shaded and not incubated between 30° and 36°C, but always incubated above 36°C. In the yellow-waffled lapwing Jayakar & Spurway (1995a, b) found that a nest with eggs was often left unattended after the sun rose but that, as

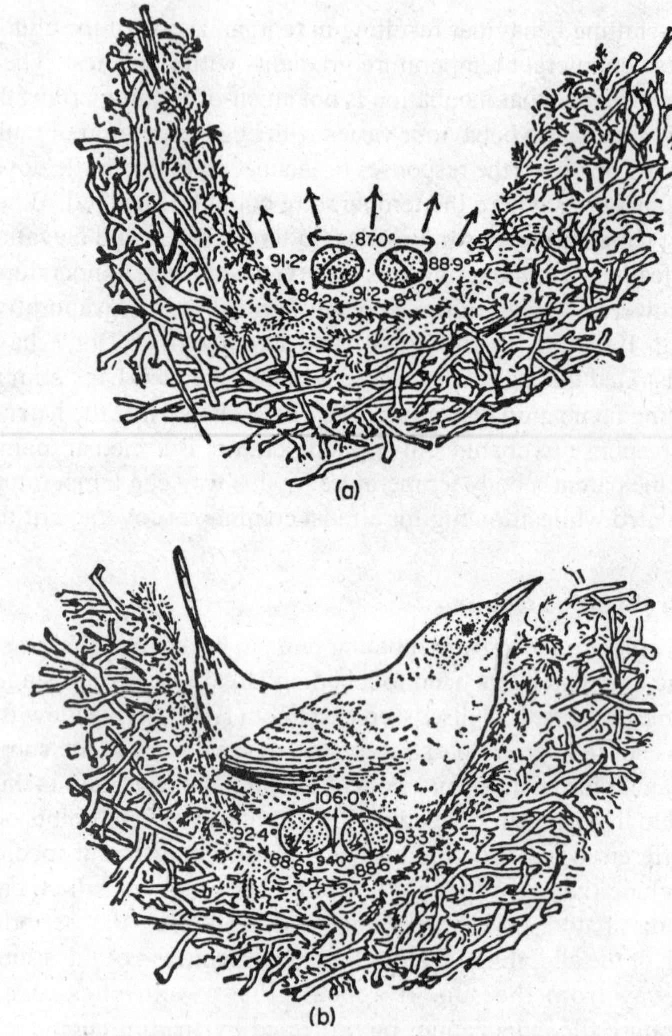

Figure 3.1 : Temperature gradients in the nest of an eastern house wren during a recess (above), and during a session (below).

the day became hotter, it was attended for longer and longer periods until it was covered almost continuously. On cloudy days periods on the nest were substantially reduced. Although the bird on duty never 'shaded' the eggs from the sun (by standing over them) its behaviour suggested that it was cooling rather than heating them.

The Continuity of Attentiveness

Under artificial conditions attempts are made to maintain incubator temperatures within quite narrow limits throughout the incubation period and this aim is consistent with high hatchability. However, in the wild, where hatching success is also high incubation temperatures are less constant. Fluctuations in nest and egg temperatures are found even in species where one or the other parent is on duty at the nest for at least 95% of the time. For example, Drent (1992) has demonstrated a tendency for the egg temperature of the herring gull to vary slightly with air temperature even when the parent is incubating. In the pigeon guillemot (which also incubates for about 95% of the time) Drent (1995) found

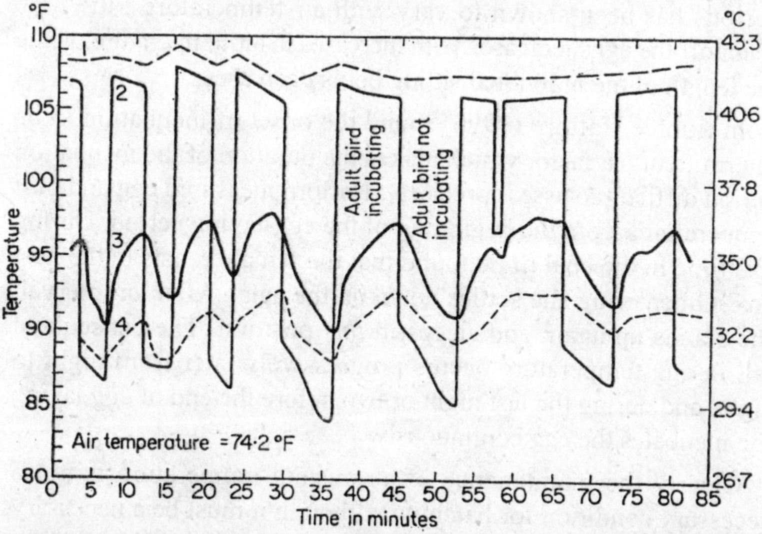

Figure 3.2 : Natural fluctuations in temperature of the eastern house wren's egg under normal conditions in the nest. (1) Body temperature of adult bird; (2) temperature at top of nest just above the eggs; (3) egg temperature; (4) temperature at bottom of nest beneath the eggs.

marked, although fairly short-lived, variations in egg temperature not only when the parents change 'shifts' on the eggs, but also within shifts, which are interrupted for short periods for defaecation, preening or drinking. Beer (1991) reports that the blackheaded gull rises and resettles on the eggs frequently during incubation bouts (on the average once every 11·3 min), and similar behaviour may be observed in the guillemot, a species where the parents take turns in incubation, and the egg is never left.

In a large number of species incubation takes place for only part of the time. During off duty periods in the wren the nest and egg temperatures fall quite considerably. In such species also there must be temperature fluctuations during attentive periods when at rather frequent intervals the sitting bird rises, shuffles about and resettles. In the brush turkey Baltin (1999) found the temperature to drop sharply when the male excavates 'control' shafts in the incubation mound. After dropping, it returns rather slowly to its normal level.

In species which incubate intermittently the length of attentive periods has been shown to vary with air temperature : the time spent off the eggs increases with increases in air temperature, while the length of the individual sitting bouts decreases.

In titmice Haftorn (1996) found the onset of incubation to be intermittent – a factor which makes the duration of the incubation period difficult to assess precisely. Haftorn measured egg and nest temperatures from the beginning of the egg-laying period. During this time in the coal tit he found that the female roosts in the nest box. On entering she settles down on the eggs. After an interval, she stands up again and sleeps in *this* position. The consequent fall in egg temperature occurs progressively later from night to night, and during the last night or two before the end of *egg* laying she incubates the egg continuously.

Clearly the maintenance of a constant temperature is not a necessary condition for hatching, although it must be a necessary condition for a constant, or a minimal, incubation period. One problem not yet settled, is whether rather frequent small temperature changes are of any significance in incubation. Russian work summarized by Lundy (1999), has claimed that fluctuations in incubator temperatures provide an increase in hatchability in the fowl; it could be of some interest to investigate these fluctuations further. We do not know whether they arise in the wild simply as a compromise between the different requirements of the eggs and of the parents, different species adapting in slightly different ways to achieve this compromise, or whether such fluctuations stimulate or assist embryonic development.

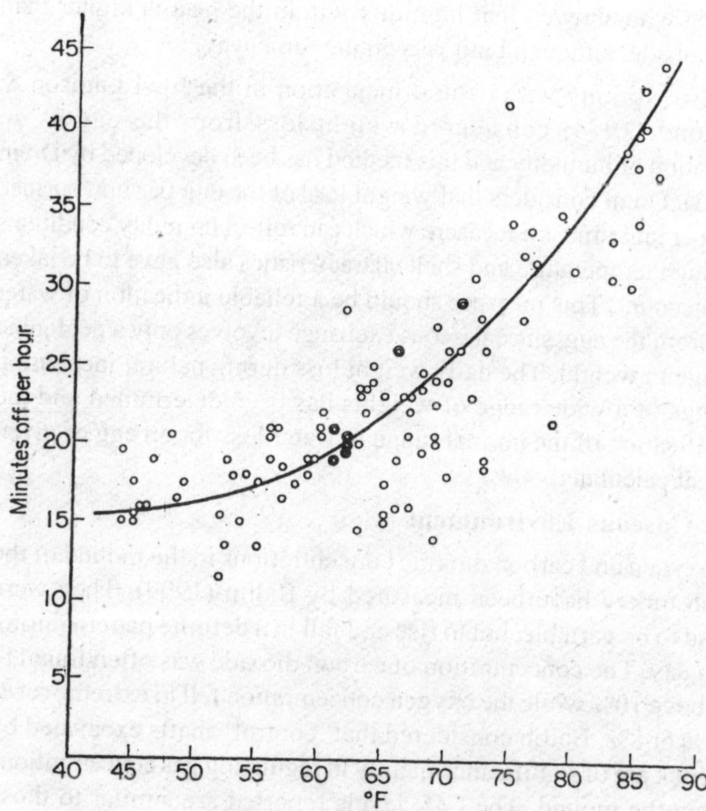

Figure 3.3 : The correlation between air temperature and time spent off the eggs.

Humidity

It has already been noted that humidity requirements appear to vary between domestic species; in the nests of wild species, however, little or no attention has been paid to direct measurements of humidity during incubation. Nests where humidity has been measured have been those of the domestic fowl, incubating under natural conditions. For example, Burke (1985) sampled a steady flow of air from a perforated egg-shaped receptacle in the nest; he found that the amount of moisture given off by the hen is fairly constant throughout the incubation period. Similar measurements have been made by Koch & Steinke (1994) and by Chattock (1995) who showed that humidity within the nest is higher than that outside, although both vary in the same way.

Also working with natural incubation in the fowl Lamson & Edmond (1974) considered weight loss from the egg as an indication of humidity and this method has been developed by Drent (1993). Drent considers that weight loss of the egg per unit surface area per unit time is a measure which can reflect humidity conditions although temperature and shell characteristics also have to be taken into account. This measure should be a reliable indication of water loss from the egg, since gaseous exchange involves only a negligible change in weight. The daily weight loss during natural incubation of eggs of a wide range of weights has been determined and the specification of the normal range in water loss for an egg of given weight calculated.

The Gaseous Environment

Oxygen and carbon dioxide concentrations in the mound of the brush turkey have been measured by Baltin (1999). They were found to be variable, but to rise and fall in a definite pattern during each day. The concentration of carbon dioxide was often found to be above 10% while the oxygen concentration fell in extreme cases to 7% or 8%. Baltin considered that 'control' shafts excavated by the cock are of significance mainly in regulating gas concentrations within the mound. The CO_2 levels reported are similar to those found in the air space of the egg of the fowl during the last few days of incubation.

There is early work on the gaseous environment in nests of the domestic fowl, when the female is incubating naturally. For example Lamson & Edmond (1974) placed a perforated wooden 'egg' in the clutch of a sitting hen and sampled air from it. They found that the CO_2 concentration within the nest increases from the beginning to the end of the incubation period, reaching a final level of 3-5%. Burke (1985) obtained similar results.

Measurements made in nests of other species appear to be rare. It has, however, been suggested that the frequent rising and shuffling carried on by incubating parents might serve to 'air the nest' to some extent. In titmice (hole-nesting species which build particularly wellinsulated nests), this behaviour has been observed by Haftorn (1996). While incubating, the female great, marsh or coal tit frequently rises, stands over the eggs with head down in the nest cavity, and with the tail up. She then works over the nest lining with trembling or shaking movements of the bill before shifting the eggs and settling on them with characteristic 'resettling' movements. Baldwin & Kendeigh (1982) reported – that the incubating wren stirs about in the nest every few minutes during the day and the night. The possibility of a relationship between nest insulation and parental activity of this kind could be examined by sampling the air before and after bouts of activity.

Egg Shifting or Turning

In most nests the eggs are frequently turned, or shifted about; eggs may be rolled, moved to another part of the nest, or simply jostled. Under natural conditions turning is likely to be accompanied by a drop in temperature, by vibratory and acoustic stimulation, and (during the hours of daylight) by an increase in photostimulation.

Egg shifting has been observed in the domestic fowl, when incubating. Eycleshymer (1987) made nests with felt sides and glass bottoms; by marking the eggs he found that they were turned more than 5 times a day. Chattock (1985) registered the position of each egg in a number of nests of the fowl. He found that no hen failed to shift her eggs very thoroughly to all parts of the nest, and intervals between bouts of turning varied from about 10 to 55 minutes.

Among wild species herring gull and blackheaded gullnests have been observed from below through glass panels. Drent found that the parent rose and resettled on the eggs every 20 minutes or so and that the eggs were moved once in about 7 resettlings. Beer found a resettling rate of about four per hour, mostly without moving the eggs. In nests of titmice observed by Haftorn (1996) eggs were shifted every few minutes.

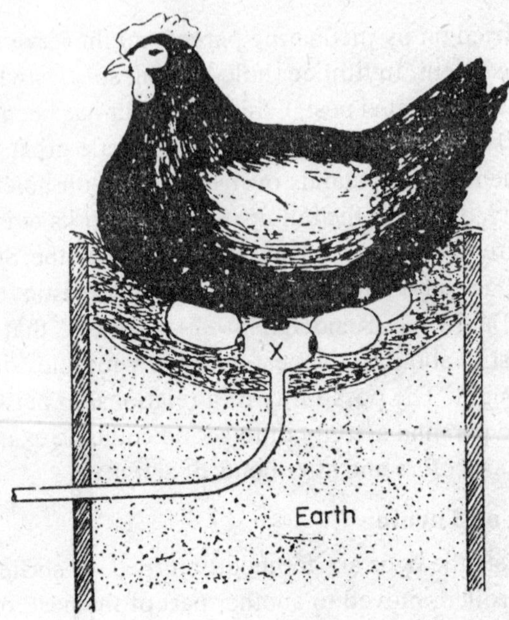

Figure 3.4 : Method of drawing a sample of air from under a sitting hen. X = a perforated wooden egg.

Tinbergen (1983) pooled the data from several herring gull nests and found that changes in egg position were irregular, or largely random. Since then Lind (1991) has considered factors underlying egg position. In the black-tailed godwit he found this to be determined partly by the frictional resistance of the nest, and partly by the egg's centre of gravity, which changes in the course of incubation. For the first few days the centre of gravity is situated almost centrally, but after 4-6 days it changes, depending on the positions of the air space, the yolk and the embryo. Towards the end of the incubation period it becomes fixed, as the air space becomes larger and more asymmetrical. The egg can therefore be turned into any position, but towards the end of incubation it tends, after being moved, to roll back into the nest cup with the heaviest part towards the bottom. In this way the parent's rather random egg-shifting behaviour results in the eggs being turned sufficiently at early stages when this is necessary for the embryo's correct development, while later they lie predominantly with the lighter side upwards and pip at the top. The embryo is then in the 'back up' position – a position which facilitates normal breathing.

According to Poulsen (1983) egg turning occurs in nearly all species. Exceptions include the palm swift which builds on the undersides of palm leaves, glues its eggs to the nest and broods them vertically, and also megapodes, where the eggs are buried upright in the incubation mound, and remain in this position until hatching. In the swift it would seem that the egg contents could be shifted by leaf movements. In megapodes Baltin (1999) has observed that the air space is movable. Possibly this, or other structural peculiarities, obviates the necessity for turning in these species.

Hatching Success

Hatchability in incubators cannot be compared directly with that found in the wild. However, when losses due to bad weather conditions, food shortages (which can result in parents deserting their nests), predation and infertility are taken out, hatchability in the wild is often high. For example in: bobwhite quail living in semi-wild conditions Stoddard (1981) found a high percentage, varying from 85% to 93% over a number of years. In the herring gull Drent (1990) reports a hatching success of about 85% of fertile eggs not taken by predators. Gulls, however, fly up from their nests in response to disturbance, a factor which is likely to reduce hatchability. In the relatively undisturbed colony of gannets on the Bass Rock, Nelson (1996) found that about 82% of eggs hatched, about 7·3% were infertile and about 11% were 'lost' mainly on account of aberrant parental behaviour, or through an accident (eggs left unattended were usually taken by gulls).

In song birds,, where the nests may be less vulnerable, hatching success appears to be slightly higher. A survey of nesting titmice from 1987-1993 has been summarized by Lack (1985): excluding nests robbed by a variety of predators about 90% of eggs of the great, blue and coal tit hatched. Seel (1998) found 92% of house sparrow eggs and 96% of tree sparrow eggs to be fertile. Omitting losses due to desertion or destruction, but including infertile eggs, hatchability was 88% and 93% respectively.

Lack (1996) discusses hatching success in a number of species, the most successful of these being the blackbird where Snow (1988) reported that about 90% of eggs hatched in the Botanic Garden in Oxford (small territories) and between 92% and 95% in Wytham woods (larger territories). In the Botanic Garden Snow related hatching success with parental age where both parents were more than one year old 95% of eggs hatched. Failure to hatch was nearly always due to infertility.

4

Early Development

The evolution of land-laid eggs marked a great advance in the history of life on earth. It is likely that the first land-laid vertebrate eggs were those of amphibious ancestors of the reptiles, which lived in water and fed on fish. There were advantages in thus placing the eggs on land, for the embryos, unlike water larvae, were not subjected to the fouling of the water and a lack of oxygen. Neither were they in danger of being consumed by the voracious predators which inhabited the ponds.

But there were also great hazards to be met and problems to be solved in laying eggs on land. The newly hatched young had to be well enough developed to crawl on land and fend for themselves; and this required a large store of nutrient in the egg, in the form of yolk, and an extended embryonic period. The eggs and embryos also had to endure dryness and the harshness of rugged surroundings; and for this purpose they were supplied with a watery envelope of albumen, tough outer membranes, and often a hard shell. Furthermore the embryos needed special adaptations to protect their delicate outer surfaces and to carry on respiratory exchange of gases with the surrounding air. They needed receptacles to retain the wastes which, if they accumulated within the embryo proper, would poison them. To serve these several functions the embryo developed four *fetal membranes,* namely, the chorion, amnion, yolk sac, and allantois. A reorganization of the circulation was also needed, for gills could no longer be used for respiration. Finally there was evolved a method of excretion whereby wastes could be eliminated with the least possible loss of water and useful substances. This was accomplished by substituting a relatively insoluble waste, uric acid, for the highly soluble wastes, ammonia and urea. Eggs which meet these specifications are known as enclosed or *cleidoic eggs* and are characteristic of reptiles, birds and the most primitive mammals, the monotremes.

The hen's egg has long been the favorite cleidoic egg for embryological study. Its earlier stages are similar to those of mammals, even of man himself. It is easy to procure-or was easy until unfertilized eggs dominated the market. Aristotle's naked-eye observations were made on the developing chick. Malpighi and Buffon formed their opinions of preformation on what they saw-or thought they saw-in opened chick eggs. Wolff and von Baer found in the chick convincing evidence that the preformation doctrine was in error. All these, however, overlooked the earliest stages in development, namely, those which take place within the body of the mother hen before the egg is laid and which therefore can be studied only by killing the hen.

THE FORMATION OF THE HEN'S EGG

The Ovarian Egg or Oocyte

The development of the chick embryo begins in the ovary of the mother when she is herself an unhatched chick. It begins when an oogonium or egg-to-be commences to transform into an oocyte or unripe egg. The microscopic oogonium becomes surrounded by a jacket of cuboidal follicle cells, probably sister cells (or rather cousins) of the oogonium; and with their help it grows, at first gradually, forming within it a central sphere of white yolk. Then, when the time comes for the oocyte to ripen, layer upon layer of yellow yolk alternating with white are added in a daily cycle until the cell-the "yolk" of the egg, as we commonly know it—has become a sphere an inch and a half or so in diameter. In six days it increases its mass as much as two hundredfold. It seems obvious that all this substance could not be synthesized and deposited without the help of other cells, and indeed it is not. Biochemical analyses confirmed by serological studies have revealed that the proteins of the egg yolk are synthesized in the mother's liver, circulated to the ovary, and deposited as yolk fluid and granules. Just what part the follicle cells play is not entirely clear.

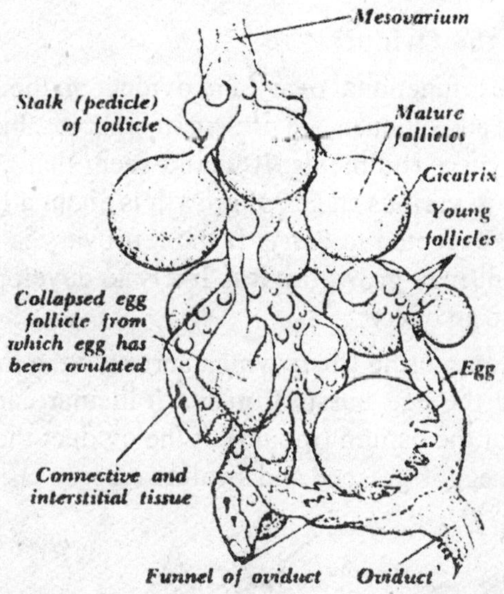

Figure 4.1 : The development of oocytes in the ovary of the hen.

The nuclei of the youngest oocytes are at the center of each cell; but as each oocyte grows in size, its nucleus moves from the center, keeping close to one side, namely, the animal pole. The cytoplasm is at first distributed throughout the oocyte; but as growth takes place it becomes more and more limited to a "germinal disc" or *blastodisc* at the animal pole. At the center of the disc lies the nucleus. For a time a thin layer of cytoplasm remains also at the periphery of the cell, but ultimately it disappears except for the disc. Beneath the disc and extending to the center of the oocyte is a column of white yolk known as the *latebra.*

A connective tissue sheath (theca) develops around the entire follicle and is well supplied with blood vessels except at the pole opposite to the attachment to the ovary. When the follicle has reached its full size, a membrane, the *zona radiata,* forms between the oocyte and surrounding follicle cells. At ovulation the sheath ruptures, and the ovum surrounded by the zona radiata, now known as the *vitelline membrane,* enters the body cavity. Normally the ovary of a hen contains follicles with their oocytes at various stages of development. The youngest are microscopic; the oldest are full-sized yolks ready to ovulate.

The Egg in the Oviduct

Birds have a functional ovary and oviduct on the left side only. Right ovaries and oviducts are present in embryos, but they regress and become mere rudiments. It would seem that a single ovary with its eggs in various stages of growth is about all that a bird of flight can be expected to carry. If the left ovary is removed, the right rudiment may grow, but it is likely to develop into a testis instead of into an ovary.

At ovulation muscle fibers which surround the mature follicle contract, and the egg bursts from its follicular capsule. Active movements of the ostium (mouth) of the oviduct then take place, by which the egg is grasped and swallowed.

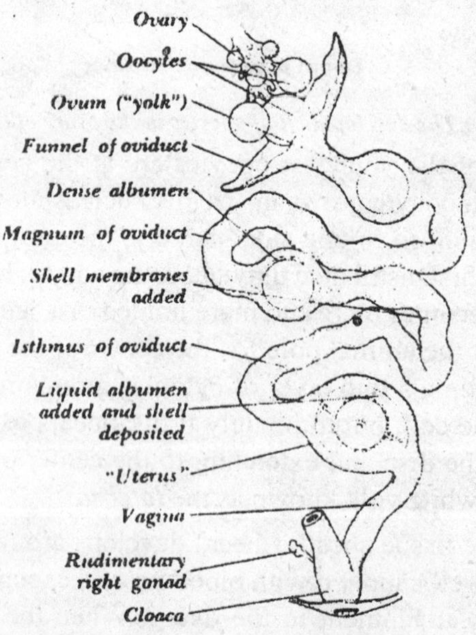

Figure 4.2 : The egg in the oviduct of the hen. Normally only one egg descends the oviduct at a time.

As the egg, the so-called yolk, passes down the 'first or convoluted part of the oviduct (magnum), a thick layer of dense albumen is secreted around it and extends as cords (chalazae) before and behind it. The passage takes about three hours. The egg next spends one hour or so in the short isthmus which follows the convoluted part of the oviduct. While here two shell membranes are secreted around the albumen. Following this the egg with its membranes enters the "uterus," where a watery fluid soaks through the shell membranes and accumulates around and within the investment of denser albumen. This is possibly the result of osmosis, for the dense albumen has a slightly higher osmotic pressure than the more liquid albumen. Thus the egg becomes distended and takes the form of an oval. For 18 or 20 hours it remains in the uterus, during which time a calcareous shell is formed outside the shell membranes.

The albumen and egg membranes rotate during the period the egg is in the uterus. This is a result of muscle contractions in the uterine wall; but the egg proper, that is, the yolk, now free within the albumen sac, does not rotate. The explanation is that the cytoplasm of the blastodisc is lighter in weight than the rest of the yolk and as a result floats on top. As a consequence of this rotation, the chalazae become spirally twisted, as anyone who opens an egg will observe.

Finally the completed egg as we know it enters the vagina, where it remains for a somewhat variable length of time and is then laid.

The rotation of the egg membranes in the uterus explains the orientation of the axes of the future embryo. According to von Baer's rule (1828), the axis of the embryo is crosswise to the long axis of the egg. The sharp end of the egg is on the embryo's right; the blunt end is on its left. However, there are exceptions. There is a correlation between the direction of coiling of the chalazae and the orientation of the embryo. The head is oriented in the direction toward which the membranes rotated. By removing eggs from the uterus 10 to 12 hours before it is time for them to be laid, and rotating them artificially, Vintemberger and Clavert were able to control the orientation of the embryonic axis at will.

DEVELOPMENT PRIOR TO LAYING

Maturation and Fertilization

As usual among the vertebrates, the first polar body is given off about the time that the ovum bursts from the ovary. Maturation proceeds as far as the metaphase of the second meiotic division, then pauses and awaits the entrance of the fertilizing sperm.

In the hen, the sperm, if present, will have migrated the full length of the oviduct and will be found in great numbers in the ostium of the oviduct. Since a hen may lay fertile eggs for as long as three weeks after insemination, it is evident that the sperm have long lives. Normally several sperm enter the blastodics as the ovum moves into the ostium. Only one, however, becomes the male pronucleus which combines with the egg pronucleus. The other sperm nuclei mndergo abortive attempts at cell division and then disappear. After the sperm has entered the ovum, a second polar body is given off. Then the egg and sperm pronuclei come together and become the zygotic nucleus.

Cleavage

The cleavage of the chick egg is incomplete, i.e., meroblastic. Only the blastodisc at the animal pole subdivides' into cells. The first four or five divisions consist of vertical furrows which do not cut very far inward and which do not separate the cells from the underlying yolk. After the fifth cleavage, however, horizontal cleavage planes undercut the central cells. The cytoplasm which remains beneath these cells then disappears so that the disc of cells, now rightly termed the *blastoderm,* is separated from the underlying yolk by a shallow *subgerminal cavity.* Incompletely separated cells and scattered nuclei remain for a time around and beneath the perimeter of the blastoderm, but soon they disappear.

The Blastula

Cleavage continues as long as the egg remains within the body of the hen. For a time the blastoderm is a delicate layer of rounded and somewhat loose cells a little over 3 mm in diameter and four or five cells thick. The cells of the periphery are somewhat larger and contain more yolk than the central cells. Moreover, they adhere to the underlying yolk. This outer zone is known as the *area opaca*. The cells of the central area which overlie the subgerminal cavity, on the other hand, are smaller and relatively free from yolk. They constitute the *areapellucida.* It has been asserted that at this stage the cells of the pellucid area are of two kinds which differ in size and yolk content and which are intermingled. This interpretation has been challenged.

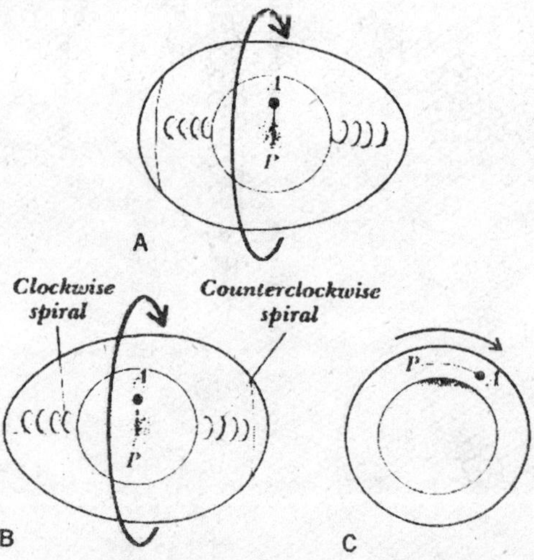

Figure 4.3 : The rotation of the egg in the hen's uterus determines the orientation of the embryo. The arrows indicate the direction in which the envelopes of the egg revolve. The egg proper ("yolk") does not revolve. As a result the chalazae become coiled, clockwise on the future embryo's left, counter-clockwise on its right. The embryo's head is in the direction toward which the egg envelopes revolve. A. The orientation is usually according to von Baer's rule: the blunt end of the egg and the clockwise chalaza are on the embryo's left. B. Sometimes the sharp end of the egg is on the embryo's left. C. View from the future embryo's right side.

The nature of the subgerminal cavity also has been a matter of dispute. In the past it has usually been considered to be a blastocoel, homologous with the blastocoel of an amphibian blastula. However, it originates by a disintegration of the cytoplasm and yolk beneath the central cells. At no time is it actually a cleft between cells as it is in the amphibian blastula. Soon, however, a true cleft does appear in the midst of the cells of the blastoderm. Some authors consider this to be the true blastocoel. The cells of the pellucid

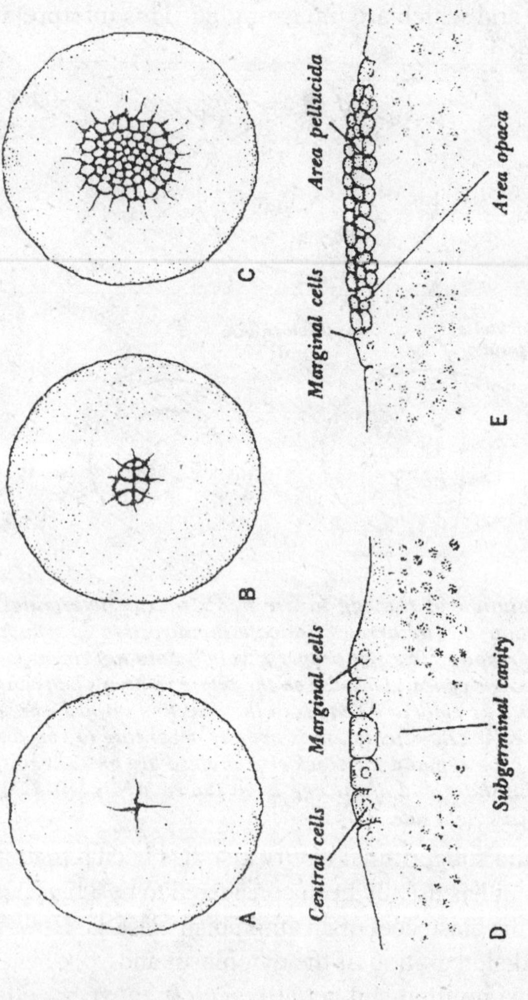

Figure 4.4 : Meroblastic cleavage in the hen's egg. A. Blastodisc following the second cleavage. B. Blastodisc following the fourth cleavage. C. Blastodisc following the seventh cleavage. D. Section through the blastopore after five cleavages. E. Section after about eleven cleavages.

area segregate. According to a study of the pigeon egg, the smaller cells which contain less yolk remain at the surface and organize into a thin outer layer known as the *epiblast.* The larger yolk-laden cells move inward one by one or in loose clusters and form a less organized understratum, the *hypoblast.* The process has been termed *delamination,* and it continues for a time even after incubation has begun. It is noteworthy that this splitting away of the hypoblast from the epiblast takes place first and principally in the posterior quadrant of the pellucid area.

While delamination is going on, the margins of the blastoderm continue to grow outwardly across the surface of the yolk. This results from active cell proliferation and also from a thinning of the blastoderm. In part it is an actual stretching of the blastoderm due to marginal cells of the area opaca which crawl outward in amoeboid fashion across the surface of the vitelline membrane. Ultimately the epiblast and hypoblast each become only one cell layer thick except in the posterior quadrant. The margins of the expanding blastoderm soon overlap the yolk so that there is a narrow *margin of overgrowth* peripheral to a *zone of adhesion* between the blastoderm and the underlying yolk.

At about this stage the egg is laid, and further development does not take place until incubation has begun. If the egg is not incubated within a few days, regression sets in.

GASTRULATION

Gastrulation in the chick is the process by which the blastoderm gives origin to the three germ layers: ectoderm, chorda-mesoderm, and endoderm.

The Unincubated Egg. Stage 1

Hamburger and Hamilton have arbitrarily defined and pictured "normal stages" in the development of the chick. They designate the fertile unincubated egg as stage 1 . As interpreted by Pasteels, it is a late blastula or early gastrula. The actual degree of development varies with the time the egg is retained in the vagina of the hen before it is laid.

At this stage, a morphogenetic movement of the hypoblast is

already under way. It consists of an outward streaming in every direction from a "growth center" located in the posterior quadrant not far from the margin of the pellucid area. This streaming may be compared to the invagination of endoderm in amphibian development.

Spratt and Haas have demonstrated this streaming in experiments in which they removed blastoderms of unincubated eggs, placed them upside down on clots of an agar medium, and then marked the undersurface (now the top surface) of the hypoblast with bands of carbon or carmine grains. They then recorded the displacement of the particles which followed.

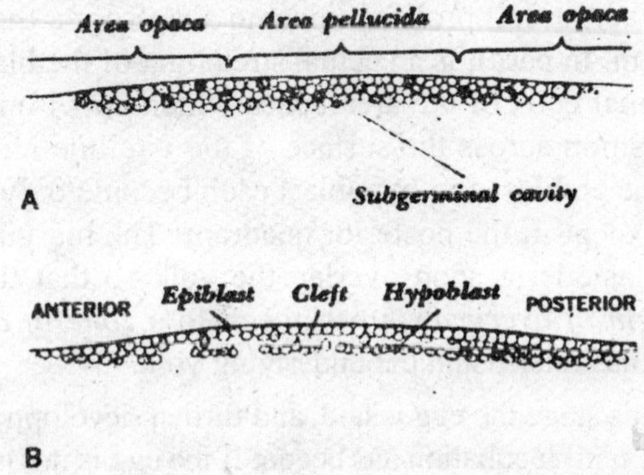

Figure 4.5 : Schematic sections through the early chick blastoderm. A. Before delamination has taken place. B. During delamination.

What is the state of determination of the unincubated egg? Is it a mosaic of self-differentiating regions, each region committed to its own particular fate? No. Quite the contrary, the unincubated egg possesses great powers of regulation, so great in fact that the term "regulation (which suggests a revision of a condition already present) hardly seems appropriate. If a blastoderm is divided into halves by a cut in any direction, twin embryos often result. Indeed Spratt and Haas have found that as little as an eighth of the pellucid area-any eighth-may form a whole and complete embryo, provided only that it includes a part of the marginal zone (namely, the zone where the area pellucida joins the area opaca). This capacity for regulation is retained through the early primitive streak stages.

If the parts of a blastoderm of a newly laid egg are capable of becoming whole embryos, why, then, does only one embryo normally form? Spratt and Haas offer the hypothesis that the hypoblast possesses a gradient pattern with respect to the density or concentration of the cells which compose it. It is thickest at the growth center of the posterior quadrant of the pellucid area. From here its density decreases in every direction, except that it is again somewhat dense in the marginal zone that borders the pellucid area. It is least dense in the central portion of the anterior half of the pellucid area. Here the hypoblast cells are at first loose and scattered. Now, according to this hypothesis, the active flow of hypoblast cells away from a denser region (normally from the growth center, but away from the marginal zone in experiments with fragments), organizes the axis of an embryo. Bilateral symmetry is therefore a consequence of the eccentric position of the growth center.

We may even have here a hint as to why the axis of the embryo is usually at right angles to the axis of the egg. It seems likely that when the egg rotates in the uterus, the outer cortical layer of the uncleaved blastodisc is dragged somewhat in the direction in which the membranes of the egg rotate. The margin of the blastodisc which trails gives rise to the growth center in the posterior quadrant. Recall that in amphibian development, also, the cortex shifts toward the side at which the sperm enters, and then the opposite side (which trails) becomes the gray crescent and later the posterior end of the embryonic axis.

The Initial Primitive Streak. Stage 2

During the first several hours of incubation, the epiblast exhibits no morphogenetic movements other than those involved in the expansion of the blastoderm. Then, at about the sixth or seventh hour, it begins to gather toward the growth center in the posterior quadrant. Its cells swing backward in circular orbits in much the fashion that a folding fan might be gathered inward toward one of its radii. At the same time the posterior radius elongates somewhat in a forward direction. By the eighth hour the cells of the posterior radius have begun to heap up and form a short, broad, longitudinal strand which is thicker than the rest of the epiblast of the pellucid area. This is the *initial primitive* streak.

Formation of the Chorda-Mesoderm

Cells from the undersurface of the epiblast of the primitive streak break away and push forward and laterally between the sheets of the epiblast and hypoblast. In Spratt's earlier experiments, he placed particles of carbon on the epiblast anterior to the streak

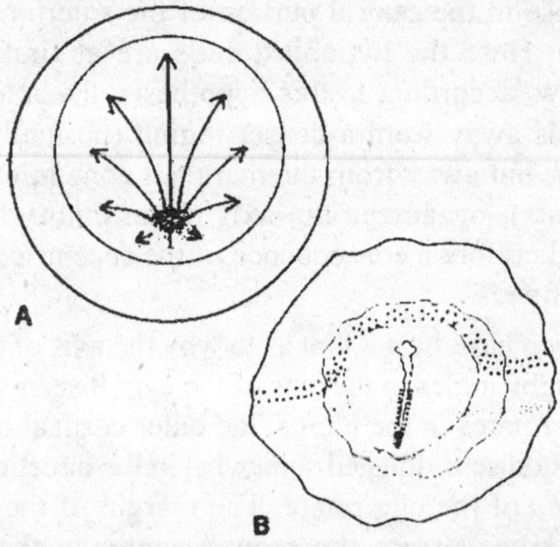

Figure 4.6 : Morphogenetic movements of the hypoblast. A. Spratt and Haas's diagram of the undersurface of the chick blastoderm at the beginning of incubation. Note the dense "growth center" in the posterior quadrant from which the hypoblast "fountains" out. B. The pattern which results when the undersurface of the unincubated blastoderm is marked with transverse bands of carbon and carmine particles and then incubated.

and found that they sank inward as soon 'as the lengthening streak reached them. This labeled epiblast gave rise to pharyngeal endoderm and notochord. (In more recent work Spratt notes that it is mainly the inner layer of the epiblast which contributes to the endoderm and mesoderm.) Epiblast material lateral to the forward end of the streak, moves toward the streak, sinks inward, and becomes axial mesoderm. Farther back it becomes lateral mesoderm. This is a true gastrulation movement, comparable to the involution of chorda-mesoderm which takes place around the amphibian blastopore and later at the tail bud.

We may anticipate a bit by noting that although there is no open blastopore in the chick, yet in ducks, reptiles, and mammals, including man, there is a tongue of cells, the so-called *head process,* which pushes forward from the anterior end of the primitive streak. It acquires a canal, the floor of which fuses with the hypoblast and then breaks down. As a result, the walls of the canal open out, and its most dorsal cells become the roof of the archenteron. Its central axis becomes the notochord. The term head process in chick embryology refers to the notochord.

The Intermediate Primitive Streak. Stage 3

The initial streak grows in length at its anterior end, partly by proliferation, but mainly by the accretion of cells which come to it from the sides. By the 12th hour the intermediate streak is about one-half the length of the circular pellucid area. It continues to grow in length, but from this stage onward it does so mainly by elongating backward at its posterior end. At the same time the area pellucida becomes oval and then pear-shaped as though to accommodate the lengthening streak.

Some forward movement within the streak itself also takes place, as a result of which the anterior end of the streak develops a thickening known as the *primitive node,* or Hensen's knot. For a time involution of chorda-mesoderm continues to be especially active just posterior to the node.

The Definitive Primitive Streak. Stage 4

By the 19th hour of incubation the primitive streak has attained its greatest length and its fullest development. It is now referred to'as the *definitive streak.* Just posterior to the primitive node is a pit, the *primitive pit.* Continuing posteriorly from the pit and extending the full length of the streak is a groove, the *primitive groove.* On each side of the groove are ridges, the *primitive ridges.* The streak ends posteriorly in a diffuse area, the *primitive plate.* The node and the ridges are the result of the heaping up of cells which have migrated in from the sides. The pit and the groove are apparently the consequences of the sinking in of cells. By the time the definitive streak stage is attained, the process of involution at the node and pit has ceased. The node has become a region of proliferation comparable to the tail bud of an amphibian. But involution of chordamesoderm still continues along the primitive ridges and groove.

At the stage of the definitive streak and immediately following it, the homologies between the chick blastoderm and the late amphibian gastrula (i.e., with a slit blastopore) are clear:

1. The primitive node corresponds to the dorsal lip of the closed blastopore (future tail bud).

2. The primitive pit represents the dorsal opening of the blastopore (neurenteric canal).

3. The primitive groove and ridges are comparable to the apposed lateral lips of the closed blastopore.

4. The primitive plate may be compared with the ventral region of the blastopore, where later the anus forms.

It is obvious from the above description that the cells of the primitive streak are in continual flux. Epiblast cells come in from the sides, heap up, sink inward, and move away from the streak as chorda-mesoderm.

The Gastrula

The stage of the definitive streak (19th hour of incubation) may be said to mark the end of gastrulation and the beginning of neurulation. The fully formed gastrula consists of three germ layers: ectoderm, chorda-mesoderm, and endoderm. The ectoderm and chorda-mesoderm are in continuity along the axis of the primitive streak, just as they are at the lateral lips of an amphibian blastopore. The endoderm is also united with the mesoderm and ectoderm at the anterior end of the streak and at its posterior end. This is true also of the amphibian blastopore.

Fate Map of the Blastoderm

It should be possible to construct a fate map of the blastoderm of the chick in the manner in which Vogt drew his fate map of the amphibian early gastrula. Such a map does not mean that the various "prospective areas" are already differentiated or that they

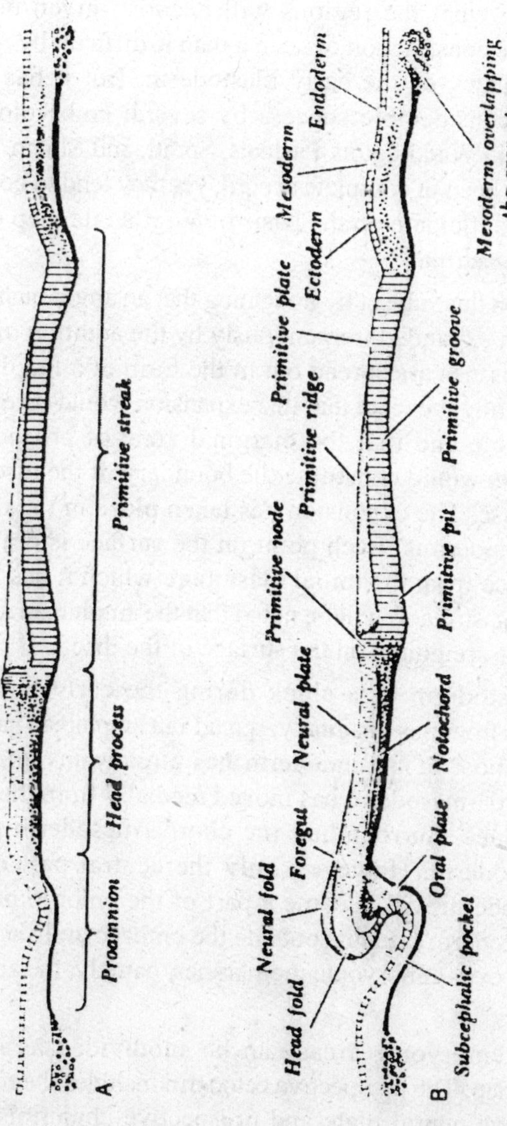

Figure 4.7 : *Median sections of the chick blastoderm. A. Head-process stage (stage 5). B. Head-fold stage (stage 6). Schematized from longitudinal sections.*

are incapable of becoming something different from their normal fate if they are isolated or transplanted to changed surroundings. It indicates only what the regions will become, given normal development. The construction of such a map is difficult by reason of the great delicacy of the early blastoderm. But it has been attempted with considerable success by several embryologists, including Rudnick, Waddington, Pasteels, Spratt, and Malan. Their findings have not been in complete accord, yet they tend to confirm what one would anticipate on the basis of Vogt's fate map of the amphibian early gastrula.

Let us approach this subject by imagining that an amphibian early gastrula has been expanded tremendously by the addition of yolk until the living tissues are spread out in the form of a flat disc or blastoderm. One might expect that this expansion would take place at the vegetal pole and that the marginal zone of prospective chorda-mesoderm would constitute the boundary of the disc. But this is not the case. The expansion has taken place in the region of prospective epidermis. Each point on the surface is drawn at that same distance from the initial blastopore which it has in the early amphibian gastrula. It will be noted that the amphibian animal and vegetal poles are points on the surface of the disc.

Now the blastoderm of a chick during the early hours of incubation differs from this imaginary, spread-out amphibian gastrula mainly in that most of its endoderm has already invaginated. Prospective chorda-mesoderm has moved medially from the sides and taken its place. Surrounding the chorda-mesoderm is the prospective ectoderm. However, only the central part of the prospective ectoderm will become a part of the embryo proper. The outlying ectoderm remains outside the embryo and shares in the formation of extra-embryonic membranes, namely, the amnion and chorion.

Each of the embryonic areas can be subdivided as in the amphibian fate map. The prospective ectoderm includes the regions of the prospective neural plate and prospective epidermis. The prospective chorda-mesoderm is divisible into prechordal meso-derm, notochord, axial mesoderm (epimere), intermediate mesoderm (mesomere), and lateral mesoderm (hypomere). The displacement of these prospective areas during the formation of the embryo.

State of the Blastoderm at the Beginning of Neurulation

We noted that when incubation begins the blastoderm is essentially undetermined. What is its state at the definitive streak stage? Have the fates of its cells been decided? If isolated or transplanted, would its parts selfdifferentiate as parts, or would they become wholes?

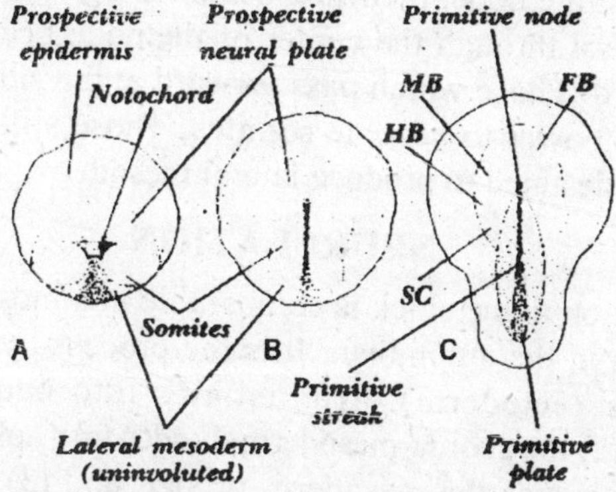

Figure 4.8 : Movements of the prospective regions of the epiblast during the formation of the embryo. A. The prospective areas at the beginning of incubation. B and C. The same areas at the intermediate and definitive streak stages. Forebrain, midbrain, hindbrain, and spïnal cord are indicated.

Many experiments have been performed in attempts to answer these questions. Rudnick cut the blastoderm into transverse bands and cultivated them on plasma clots. She found that sections taken from anterior to the primitive node give rise to neural plate, sensory structures of the head, and body wall; that is, they develop according to their expected fates. Sections which include the node produce, in addition, notochord, head mesoderm, and (laterally) heart muscle. These results are most certain if endoderm is included in the explant; less certain if it is absent. Transverse sections behind the node form no axial structures whatever-no neural plate, notochord, or axial mesoderm-whether endoderm is present or not. By the forward migration of streak cells the streak itself lengthens to form a thin, dense tongue. Sections taken from the posterior part of the pellucid area produce an abundance of immature blood cells (erythroblasts).

How are these observations to be interpreted? Experiments by Spratt, using various techniques, have established the view that the node is the continuing center of organization. The cells behind the node (corresponding to those cells in the amphibian which have not yet undergone involution) are undetermined. They are incapable, in and of themselves, of developing according to their normal fates. As these posterior cells undergo involution and move forward past the node, their fate becomes decided. Those which move forward through the center of the node become specified as notochord. Those which pass forward at the sides of the node acquire the power to become somites. Those still farther to the side are predestined to produce lateral mesoderm.

NEURULATION

Neurulation in the chick is comparable in almost every way to neurulation in the amphibian. It is the process by which (1) the outer layer (ectoderm) differentiates into neural plate and epidermis, (2) the chorda-mesoderm divides into notochord and the several regions of the mesoderm proper, and (3) the endoderm gives rise to the foregut and later to the hindgut as well. During neurulation, also, the embryo proper begins to separate from the surrounding nonembryonic (extra-embryonic) areas of the blastoderm.

Development of the Head Process. Stage 5

Neurulation may be said to begin immediately following the definitive primitive streak stage (about the 20th hour of incubation) when chorda-mesoderm cells migrate forward from the primitive node and differentiate into the central tongue of cells known as the head process. The axis of the head process becomes the notochord.

The notochord elongates by cell multiplication, stretching, and the addition of cells from the primitive streak to its posterior end.

If, as Spratt has shown, the notochordal center in the primitive node is removed at an early primitive streak stage, it may regenerate, and further development may take place normally. But if it is removed at the definitive streak stage or later, even though a normal head may form, a new notochord is not produced, and the right and left somites fuse together across the midline.

Axial mesoderm is proliferated from centers on each side of the notochordal center. If one of these centers is removed, no more somites are formed on that side. Thus the primitive node is a processing center through which the material of the primitive streak streams and, as it does so, becomes organized into notochord and axial mesoderm.

Grabowski grafted tissue taken from the primitive node into the region of the lateral mesoderm. He found that mesoderm alone, without accompanying ectoderm, is able to bring about the formation of a head. The ectoderm of the node has no such power. The mesoderm of the donor induces the ectoderm of the host to become neural tissue. Then the neural tissue, in turn, as Fraser has shown, influences the mesoderm to condense into axial mesoderm and to segment into somites.

The Regression of the Primitive Streak

As a result of the streak material moving forward through the centers of the node, the embryonic region in front of the node increases in length, and the primitive streak decreases. The process is termed the regression of the streak. In effect, the node migrates backward. Spratt states that the cells of the top surface of the node regress with the node; but for the most part the cells of the streak feed through the regressing node, so that the node is more like a moving wave than a material object.

As the node regresses, material at the posterior end of the streak disperses until finally about all that is left of the streak is the node itself, situated far back at the posterior end of the pellucid area. Its fate is to become the tail bud of the embryo, from which for many hours the neural tube, notochord, and axial mesoderm (somites of the tail) continue to be proliferated.

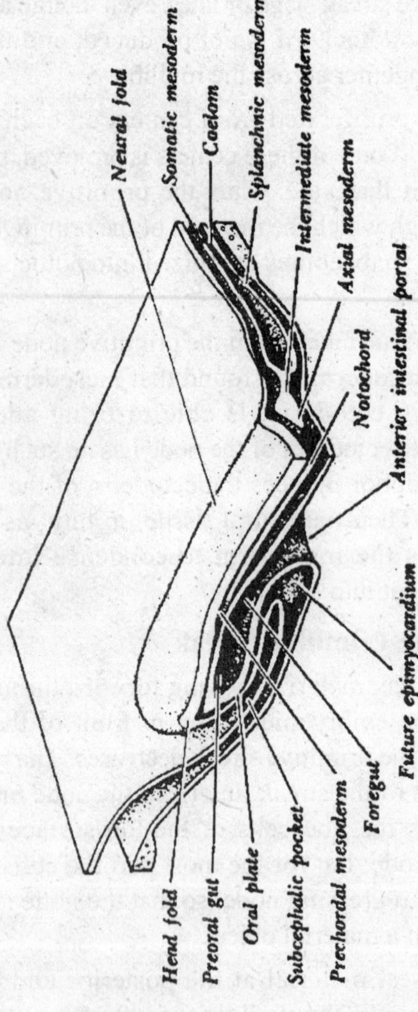

Figure 4.9 : *Stereogram to illustrate neurulation in the chick. Note the head fold, foregut, neural folds, and the derivatives of the chorda-mesoderm.*

The Neural Plate

From about the 20th hour of incubation onward, the neural plate can be recognized in stained whole mounts as a denser region of ectoderm anterior and lateral to the head process. Posteriorly it fades out opposite the middle of the primitive streak.

During neurulation the central axis of the neural plate, namely, that part which overlies the notochord, sinks inward as the neural groove, while the margins of the neural plate, i.e., the neural folds, rise up and move toward the midline, just as they do in amphibians.

The Chorda-Mesoderm

While the neural plate is folding into the neural tube, the chorda-mesoderm is also differentiating. Its most anterior part, the prechordal mesoderm, gives rise to much of the mesenchyme of the head. Behind it is the notochord. On each side of the notochord are the three longitudinal strands of mesoderm: (1) The thicker medial strands are axial mesoderm, or epimeres, which merge anteriorly into the mesenchyme of the head. (2) Lateral to the axial mesoderm are strands of intermediate mesoderm, or mesomeres. At this early stage they are thin and difficult to recognize. (3) Farthest to the side is the lateral mesoderm, or hypomeres. Only a small part of the lateral mesoderm, however, is destined to become a part of the embryo proper. Its outlying regions remain extra-embryonic and contribute to the fetal membranes. Early in development the lateral mesoderm splits into somatic and splanchnic layers, with the body cavity (coelom) between.

The Head Fold and Foregut. Stage 6

While the neural plate is folding into the neural tube and the chorda-mesoderm is dividing into the notochord and the several regions of mesoderm, the embryo proper pinches away from the extra-embryonic blastoderm. The process starts anterior to the neural plate when the embryo proper pushes forward. This results in a transverse fold, namely, the *head fold*. The pocket of ectoderm beneath the head fold is the *subcephalic pocket*.

From the first the head fold includes endoderm as well as ectoderm. The pocket of endoderm within it is the *foregut*. Its opening posteriorly onto the surface of the underlying yolk is termed the *anterior intestinal portal.*

As the lateral neural folds rise up and join to form the neural tube, they also stretch longitudinally. The result is that the head fold increasingly bulges forward and the subcephalic pocket becomes deeply recessed. The endoderm is not involved in this process. It is true that the foregut lengthens, but it does so in a different manner and in the opposite direction from the lengthening of the head fold. It is added to at its posterior end. The endoderm and accompanying splanchnic mesoderm close in from the sides beneath the foregut, so that the anterior intestinal portal is made to retreat toward the rear.

Meanwhile the blastoderm surrounding the embryo continues to grow outwardly over the surface of the yolk. Its outermost zone consists of ectoderm and endoderm only. Since the cells of the latter are laden with yolk granules, this zone is known as the *area vitellina*. Mesoderm pushes laterally and posteriorly from the embryo a part of the way into the space between the ectoderm and endoderm. This zone containing mesoderm appears mottled in stained whole mounts because of clusters of cells which contain hemoglobin. These clusters are known as *blood islands* and are located between the splanchnic layer of mesoderm and the endoderm. Soon they organize into blood vessels and blood cells, as will be described in the next chapter. Hence the zone of mesoderm is called the *area vasculosa*. For a time, mesoderm does not invade a region beneath and anterior to the head, known as the *proamnion*.

We arbitrarily close this chapter at about the 21st hour of incubation (stage 6), when the head fold has formed and the first somites have just begun to appear.

5

The Young Chick Embryo

In a broad sense any stage of development from fertilization to hatching is an embryo; but we shall use the word in a narrow sense to refer to those stages during which the main features of the new organism take form.

The "normal stages" in the development of the chick which Hamburger and Hamilton defined have served a useful purpose especially to experimental embryologists. Unfortunately their stages do not represent equal intervals in the course of development. Stating the age of the embryo in hours of incubation, as is usually done, also is not exact; for chick eggs vary in the amount of development which takes place before they are laid, and variation in the temperature of incubation makes a very considerable difference. From the 21st hour to the 40th hour it is convenient to indicate the stages of development by the number of pairs of somites. After the 40th hour somites are formed at less frequent intervals, and they are difficult to count.

THE LINEAR EMBRYO

For a time, namely, until about the 36th hour of incubation, the chick embryo is linear, and except for the heart it is symmetrical. After about the 36th hour it begins to twist and curl in a manner presently to be described. As a linear embryo it is comparable to the young amphibian embryo. However, the presence of the huge mass of yolk accounts for significant differences.

The Neural Tube

The right and left neural folds come into contact with each other at about the 5-somite stage in the region which is to become the midbrain. The contact gradually progresses forward and back, but for a time (until about the 13-somite stage) an opening remains at the anterior end of the neural tube known as the *anterior neuropore*. At the posterior end the neural folds remain open for a considerably longer time. Later, because of its shape, this opening is called the *sinus rhomboidalis*. When it does close (at about 20 somites) the neural folds enclose the primitive node and anterior part of the primitive streak.

The progressive closure of the neural tube is readily observed in living chick embryos which have been removed from the yolk, floated into a watchglass of warm salt solution, and observed by reflected light. Embryos which have been fixed, stained, cleared, mounted whole on glass slides, and then viewed by transmitted light show the sidewalls of the neural tube as dark longitudinal bands. By studying mounted cross sections of a 24-hour chick embryo, beginning at the posterior end and working forward, one gets an idea of the progressive closure of the neural folds.

The anterior half of the neural tube becomes the brain. It shows the same three divisions—*forebrain, midbrain,* and *hindbrain-which* were described in connection with the amphibian embryo. The hindbrain tapers gradually into the *spinal cord.* The early bulging of the *optic vesicles* from the sides of the forebrain is a notable feature of chick development. By the 10-somite stage they reach the full width of the head and appear to be adherent to the epidermis. A broad downpocketing of the forebrain below the vesicles is the *embryonic infundibulum.*

Almost from the first the neural tube presents a series of lateral swellings of the brain known as *neuromeres.* Their exact number and their relation to the subdivisions of the brain is a matter of controversy. Do they represent a primitive segmentation of the vertebrate brain? Soon they reach their maximum development and disappear.

The Notochord

The -notochord as seen in whole mounts of linear embryos is a narrow median strand of cells underlying the neural tube. It is entirely too weak to function as a skeletal structure. Nevertheless, it has an important morphogenic function, for around it forms the axial skeleton. If it is removed by experimental surgery, the somites of the two sides may unite across the midline.

The notochord begins at the forward end of the hindbrain and extends back to the *notochordal bulb* on the underside of the primitive node. It elongates at its anterior end until it comes close to the infundibulum on the undersurface of the forebrain. However, it is difficult at this early stage of development to distinguish the most anterior notochordal cells from the prechordal mesoderm anterior to it.

The Foregut

We noted that at first the foregut extends to the forward tip of the head fold; but the neural tube bulges forward and soon overshadows it. Its growth in length is at its posterior end and is accomplished by the closing in of the walls of the anterior intestinal portal beneath it. As a consequence of this concrescence the anterior intestinal portal recedes posteriorly. The process is readily seen when a living blastoderm is removed to a watchglass of warm salt solution, turned ventral side up, and observed by reflected light. In lightly stained whole mounts the margins of the foregut and portal appear as delicate double lines. One line is the endoderm, the other is the accompanying splanchnic mesoderm.

Near its anterior end, the floor of the foregut (endoderm) is in contact with ectoderm of the roof of the subcephalic pocket. This is the location of the oral plate. The blind extension of the foregut anterior to the oral plate is the *preoral gut* or Seessel's pocket. It is temporary.

The Mesoderm

Mesoderm is present between ectoderm and endoderm throughout the embryo, but at first it is not abundant within the head fold. Here it consists of mesenchyme, that is, loose cells derived from prechordal mesoderm and from the forward extension of axial mesoderm, together with some contribution from the neural crest.

The axial mesoderm is seen in whole mounts as dark bands on each side of the neural tube. At about the 21st hour of incubation a cleft appears across each band a short distance anterior to the primitive node. It marks the division between the 1st and 2d somites. Approximately one hour later a second cleft appears posterior to the first and divides the 2d somite from the future 3d somite. The process repeats itself approximately once an hour for 20 hours. After that the somites continue to appear at a somewhat slower rate. The most posterior somites are, of course, the youngest.

The intermediate mesoderm undergoes little development during the period of the linear embryo. It may be recognized in stained whole mounts as the less dense region lateral to the somites and medial to the more darkly stained lateral mesoderm. In cross sections it appears as "necks" of cells connecting the somites with the lateral mesoderm.

The lateral mesoderm is present at the sides of the embryo and posterior to it. Its future is partly embryonic and partly extra-embryonic, but at this early stage the two parts merge with each other without any indication of a demarcation between them.

Beneath the head fold and anterior to it, there is an area into which mesoderm has not yet entered and which is known as the *proamnion*. The extra-embryonic lateral mesoderm pushes forward on each side of the proamnion as "horns." These can be seen in lightly stained whole mounts. Ultimately they join in front of the embryo and completely surround the proamnion.

Everywhere the lateral mesoderm splits into the two layers: *somatic mesoderm* next to the ectoderm, and *splanchnic mesoderm* next to the endoderm. The body cavity, or coelom, lies between them. Part of it will become the body cavity of the embryo; but the larger part is *extra-embryonic coelom,* or exocoel, and will remain outside the embryo.

The behavior of the splanchnic mesoderm in the region of the foregut is of special interest and importance. Here, as everywhere else, it hugs the endoderm closely. It moves in from the sides as the anterior intestinal portal shifts posteriorly. As a result, two coelomic pockets or bays are formed, one on each side of the foregut, which in time unite to become the pericardial cavity. Since at this early stage they are still widely open laterally to the extra-embryonic coelom, they are best called *amnio-cardiac vesicles*. Each is bounded anteriorly by the anteriorly pocket, dorsally by

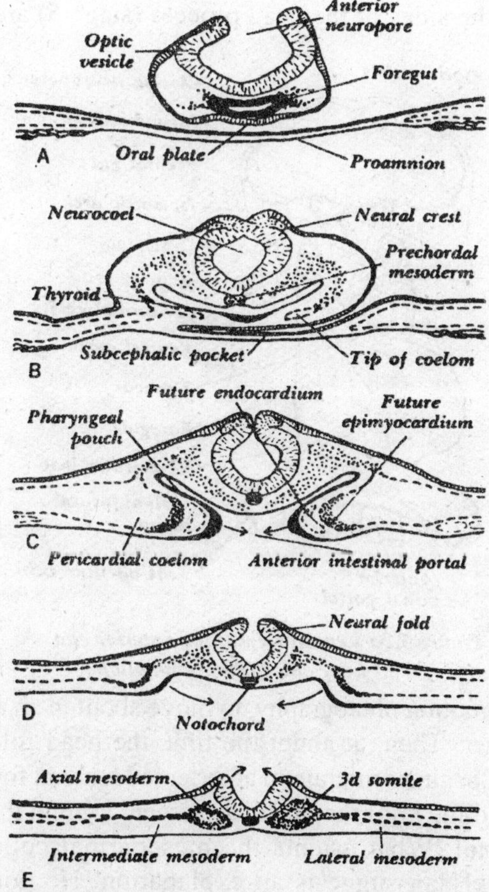

Figure 5.1 : Sections of a 24-hour chick illustrating the closure of the neural tube and foregut.

the floor of the foregut, posteriorly by the anterior intestinal portal, and ventrally by the splanchnopleure which covers the yolk, namely, the yolk sac.

According to DeHaan, this movement of the splanchnic mesoderm in the region of the foregut is not just a passive one. The mesoderm is not simply carried along with the endoderm as the latter folds in and forms the floor of the foregut. An actual migration of mesodermal cells takes place. Clusters of cells which at first are at the sides of the head process (stage 5) are seen, by

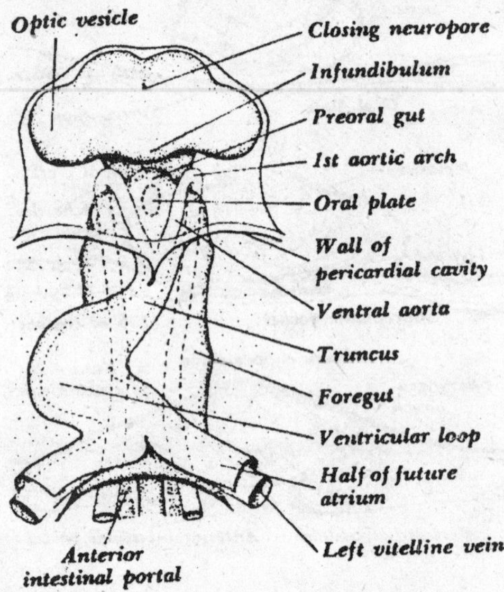

Figure 5.2 : Ventral view of the anterior end of a 33-hour chick (diagrammatic).

means of slowmotion photography, to move about in an apparently random manner. Then, at about the time the head fold appears (stage 6), they begin to migrate in an oriented fashion forward and toward the midline until they form a crescent around the anterior intestinal portal. What orients the mesodermal cells in their migration? DeHaan suggests an explanation. He notes that a change of shape of the endodermal ce!ls takes place at about the time that the migration of mesodermal clusters becomes orderly. At first the endoderm forms an irregular squamous epithelium. But at about stage 6 its cells become spindle-shaped and oriented toward the intestinal portal. He suggests that this ordered surface of the endoderm provides "contact guidance" to the migrating mesoderm. However this may be, as the anterior intestinal portal shifts posteriorly, more and more heartforming mesoderm moves into place beneath the floor of the foregut and there organizes a heart.

Extra-embryonic Blood Vessels

The lateral mesoderm spreads outwardly from the primitive streak until it reaches the area opaca. It then pushes some distance into the area opaca, and as it does so it appropriates to itself some of the yolk granules of the area, possibly even some of the cells. Where it overlaps the area opaca it forms cords of cells containing hemoglobin which are known as *blood islands*. The source of the cells is in dispute, but they may be derived from mesenchyme which migrated outwardly in advance of the lateral mesoderm.

As the blood islands extend, they unite to form a network of cords lying between the endoderm and mesoderm of the extra-embryonic splanchnopleure. Their surface cells become the endothelia of the vitelline (yolk sac) blood vessels. Their central cells become free, and most of them are red blood corpuscles known as embryonic erythroblasts. That part of the area opaca which is thus invaded and vascularized is known as the *area vasculosa*. Farther out, the area opaca remains free of mesoderm and blood vessels and is called the *area vitellina.*

A network of endothelial vessels develops also in the splanchnopleure of the area pellucida. Possibly it is the result of ingrowth of vessels toward the embryo from the area vasculosa. At any rate, the vessels soon join endothelial cords which grow outward from the embryo. Thus there is established a continuity of vessels between the embryo and the vascular region of the blastoderm.

Blood Vessels within the Embryo

The development of the blood vessels of the young amphibian embryo applies, with modifications, to the chick. It will be recalled that in the amphibian there are two pairs of primitive blood vessels, one pair dorsal to the gut and one pair ventral to the gut. Since the chick embryo is spread out flat upon the surface of a huge yolk, the picture of the blood vessels is of necessity different. The dorsal vessels are seen first in the splanchnopleure immediately lateral to the axial mesoderm. They become the *dorsal aortas* and their forward extensions to the forebrain, the *internal carotid arteries*. Posteriorly they connect with a network of vessels of the area vasculosa, the forerunners of the *vitelline arteries*.

The vessels of the chick which correspond to the primitive ventral blood vessels of the amphibian are less easily recognized. At first they are represented by the blood islands and vascular net of the area vasculosa. Blood flows forward in this net until it reaches the horns of the area vasculosa. Then it returns as the *vitelline veins* to the folds which border the anterior intestinal portal. As the splanchnic mesoderm migrates medially beneath the foregut, it seemingly pushes strands of endothelia before it. These strands become the *endocardium* of the heart and, farther forward, the *ventral aortas*. The latter pass forward on each side of the oral plate and then curve dorsally around the forward border of the foregut as the first pair of *aortic arches*.

The Heart

The layers of splanchnic mesoderm which enfold the endocardium become the muscular layer of the heart, namely, the *myocardium,* and its outer serous covering, the *epicardium*. At this early stage of development, however, the myocardium and epicardium together constitute a single layer, the *epimyocardium*. Dorsal to the heart, that is, between it and the underside of the foregut, the layers of splanchnic mesoderm meet each other and form a mesentery, the *dorsal medocardium*. This is present in the linear embryo but soon breaks through. Beneath the heart the same two layers of splanchnic mesoderm also come together, but here they break down immediately, so that a *ventral mesocardium* is at best a transitory structure. As a result, the right and left amnio-cardiac vesicles join each other both above and below the heart and form a single *pericardial cavity*. The heart is now free to twist and spiral without involving the body wall.

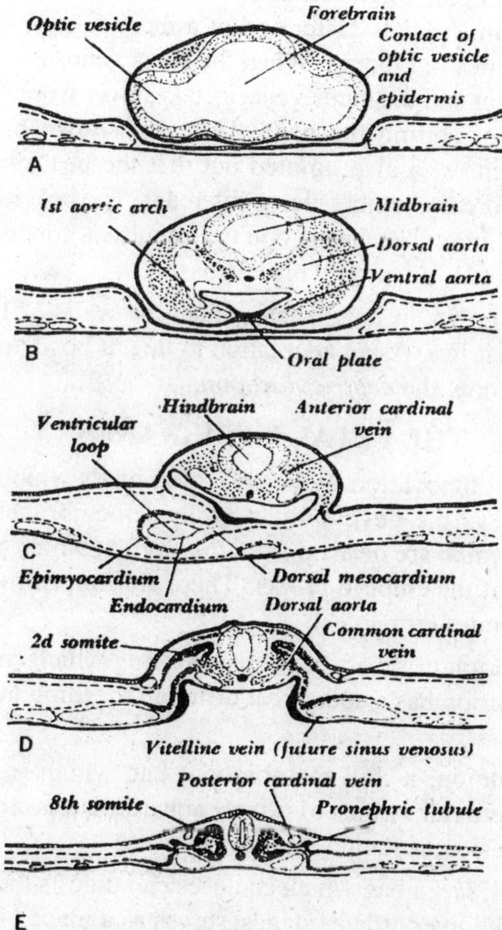

Figure 5.3 : Sections of a 33-hour chick.

The first chambers of the heart which take form beneath the foregut are the *truncus* and the anterior part of the *ventricle*. As the foregut grows in length the anterior intestinal portal recedes, and the heart is added to at the rear. At the 33d hour (13 somites) the ventricle is complete, but the atrium and sinus venosus are still represented by paired primordia in the folds of the nortal. A few hours later the atrial rudiments join in the midplane and give rise to a single *atrium;* but it is nearly two days before the primordia of the *sinus venosus* unite and form one organ.

The heart begins to twitch as soon as a part of the ventricle has taken form. It beats faster and at a more regular rate when the atrium comes into being. When the sinus venosus is added, it speeds up again and the sinus venosus takes over from the atrium the function of setting the pace. This was shown in 1920 by Florence Sabin, who also pointed out that the beat of the heart originates before nerve fibers have entered it. The beat is therefore of myogenic (muscular) origin, and the impulse is conducted from the sinus venosus forward by muscle tissue.

The heart grows in length faster than the walls of the cavity within which it lies. As an adaptation to this, it bows to the right and forms a loop, the *ventricular loop.*

THE FETAL MEMBRANES

A principal innovation in the case of animals which lay their eggs on land is the development of fetal membranes, that is, membranes which are derived from the egg but which do not become a part of the embryo proper. There are four of these in the higher vertebrates (amniotes):

1. The chorion or serosa, an outer covering which encloses all (the word chorion has a somewhat different meaning in mammal embryology)

2. The amnion, a thin membranous sac which rrounds the embryo and provides a sort of private aquarium to protect it from pressure, abrasion, irritation, and loss of water

3. The yolk sac, a bag which encloses end digests the yolk and which, like the liver at later stages, serves as a place of origin of blood cells

4. The allantois, by origin a precocious bladder, in which waste accumulates and which becomes the respiratory organ of the embryo.

Th Chorion and Amnion

These two fetal membranes are derived from the extra-embryonic somatopleure. About the 30th hour of incubation, a transverse crescent shaped thickening of the ectoderm rises up

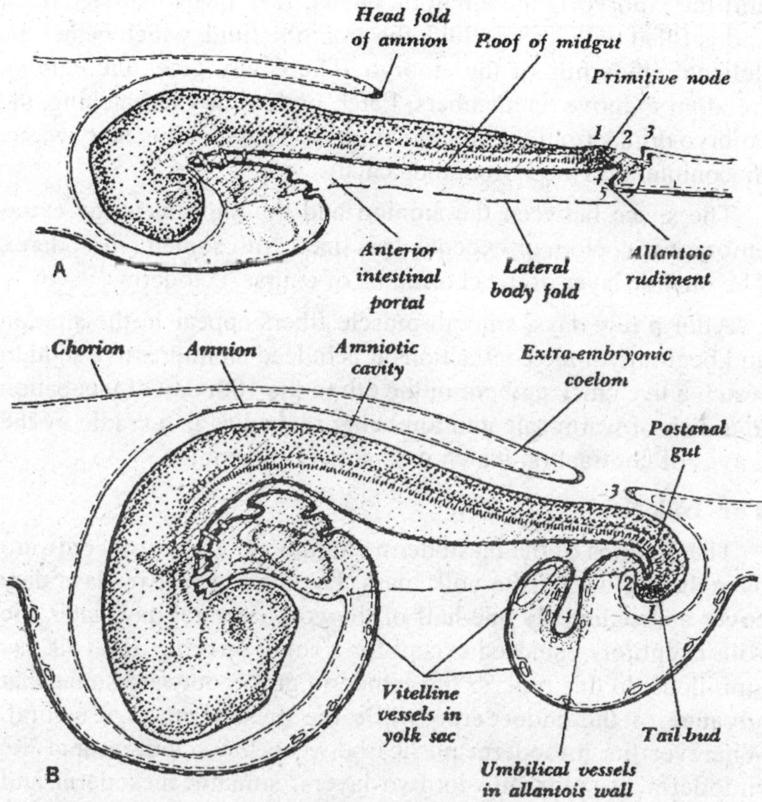

Figure 5.4 : Diagrammatic lateral views of a chick embryo illustrating the development of the fetal membranes. A. At 2 days. B. At 3 days. The embryo proper is shaded with diagonal lines. The numerals show the continuity in the posterior region.

anterior to the head and begins to move backward over the head of the embryo. It is the head fold of the amnion. At first it consists of ectoderm only, but soon mesoderm enters it from the sides. The outer limb of the amnion fold is the chorion. It has no direct connection with the embryo. The inner limb which lies next to the embryo is the amnion. The head fold of the amnion moves posteriorly over the embryo as lateral amniotic body folds move in from the sides. Tardily a tail fold of the amnion makes its appearance until ultimately all the folds come together and form a complete tent over the embryo. The space between the amnion and the embryo is the amniotic cavity. It is lined with ectoderm and is filled with a clear fluid, the amniotic fluid, which bathes the delicate epidermis of the embryo. The cavity gives the embryo freedom to move its members. Later, preliminary to hatching, the embryo drinks from the amniotic fluid and passes alimentary waste, meconium, into the surrounding cavity.

The space between the amnion and the chorion is the extra-embryonic coelom or exocoel. It is lined with somatic mesoderm. The outside layer of the chorion is, of course, ectoderm.

After a few days, smooth-muscle fibers appear in the amnion and begin rhythmic contractions It is indeed an impressive sight to watch a live chick embryo of the 6th to the 10th day of incubation in a dish of warm salt solution being rocked as in a cradle by the waves of contraction which pass across the amnion.

The Yolk Sac

The margins of the blastoderm continue to grow grow outward over the surface of the yolk mass. By the end of two days they cover approximately one-half of the yolk. Several days later, the yolk is entirely enclosed except for a small opening, the yolk sac umbilicus. In this process the ectoderm grows outward somewhat advance of the endo derm, while the mesoderm Tags behind. Wherever the mesoderm pushes between the ectoderm and the endoderm, it splits into its two layers: somatic mesoderm and splanchnic mesoderm. The vitelline blood vessels are located between the splanchnic mesoderm and endoderm.

The extra-embryonic endoderm and the splanchnic mesoderm which accompanies it constitute the *yolk sac.* The function of the endoderm is to secrete enzymes which digest the yolk and then to absorb the products of digestion into the vitelline blood vessels of the yolk sac. But the yolk sac is also the first source of blood cells and blood vessels of the embryo. Even at this early stage the vitelline blood transports oxygen and carbon dioxide. The yolk sac also performs biochemical functions which are later taken over by the liver.

The Allantois

About the third day of incubation, the region of the future floor

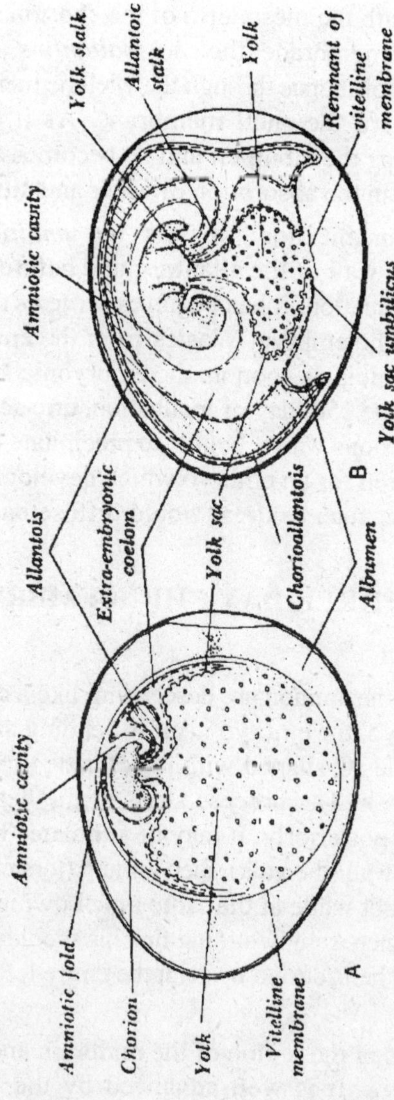

Figure 5.5 : The later development of the fetal membranes. A. At 4 days. (However at 4 days th amniotic folds are normally closed) B. At 12 days.

of the hindgut begins to bulge as a precocious bladder, namely, the allantois. The outpocketing slowly enlarges and invades the extra-embryonic coelom. Wherever the mesoderm of the outpocket comes into contact with the mesoderm of the chorion, it adheres. The result is a single membrane, the *chorioallantois.* In time the expanding chorioallantois bursts through the vitelline membrane and pushes outward toward the shell membrane. As it does so, it progressively envelops the albumen and so becomes a sac filled with albumen. It aids in the absorption of water and albumen.

A network of abundant blood vessels, the *umbilical circulation,* develops in the wall of the allantois; and before the end of the first_ week of incubation, it has begun to serve as the primary respiratory organ of the embryo. Wastes from the embryo enter the cavity of the allantois as soon as the embryonic kidneys become functional. By the 15th day of incubation, uric acid, a relatively insoluble nitrogenous waste, begins to precipitate out. This is an important adaptation for an embryo which develops in air, for a more soluble waste, such as urea, would diffuse back into the embryo and poison it.

TWO- AND THREE-DAY CHICK EMBRYOS

Flexion and Torsion

While the folds of the amnion are descending like a curtain over the head of the embryo, the embryo itself is bending and twisting so that it soon becomes C-shaped with its left side turned toward the surface of the yolk. The process starts at the head end and gradually progresses posteriorly. It may be simulated by a person standing and facing a wall (the yolk); then turning (torsion) by twisting his head to the right while at the same time bowing lower and lower (flexion); then increasingly turning first his shoulders and then his trunk until finally he is curled up with the entire left side of his body next to the wall.

The first flexure is in the region of the midbrain and is known as the *cranial flexure.* It is well advanced by the end of the second day. By this time also the head, but not the trunk, has revolved to the right. One day later strong forward bending takes place in the cervical region posterior to the hindbrain, the *cervical flexure,* and a flexure has begun at the far posterior end, the *caudal flexure*.

During the fourth day of incubation, flexion and torsion become complete.

These bends and twists make mounted cross sections of chick embryos difficult for the beginner to interpret. The student will do well to have a whole mount of the same age at hand (or a figure of the same) and to constantly refer to it for the location of the sections he is studying.

Why does the bird embryo twist and turn? Reptile and mammal embryos do the same. Is it not so that the embryo will fit compactly into the amnion or egg membrane or uterus? Not until the time of hatching (or birth) does the embryo straighten out.

The Nervous System and Sense Organs

The statements which were made concerning the nervous system of the young amphibian embryo apply also to the 2- and 3-day chick. The primary brain vesicles—forebrain, midbrain, and hindbrain—are even more clearly defined in the chick than in an amphibian of comparable age. By the 48th hour the forebrain is divided into telencephalon and diencephalon.

Olfactory placodes are present in the epidermis of the 48-hour chick at the sides of the telencephalon. By the 72d hour these placodes have shifted ventrally and have caved in to form the *olfactory pits* or vesicles. By the 72d hour, also, the walls of the telencephalon adjacent to the placodes have begun to bulge laterally and to form the *cerebral hemispheres*.

The *optic vesicles* which pushed laterally from the diencephalon have caved in and formed the *optic cups* (cupping began at about the 40th hour). They are connected with the diencephalon by narrowed *optic stalks*. It will be noted that each optic cup is incomplete on its ventral side. The cleft which is present there is known as the *choroid fissure*. It permits blood vessels to enter and leave the cavity of the cup.

Lenses also make their appearance at about the 40th hour as thickenings of the epidermis where the epidermis is in contact with the expanded optic vesicles. By the 48th hour each lens placode has sunk inward and has become a pit open to the exterior. By the 72d hour it has separated from the epidermis and has formed a vesicle which occupies the center of the optic cup.

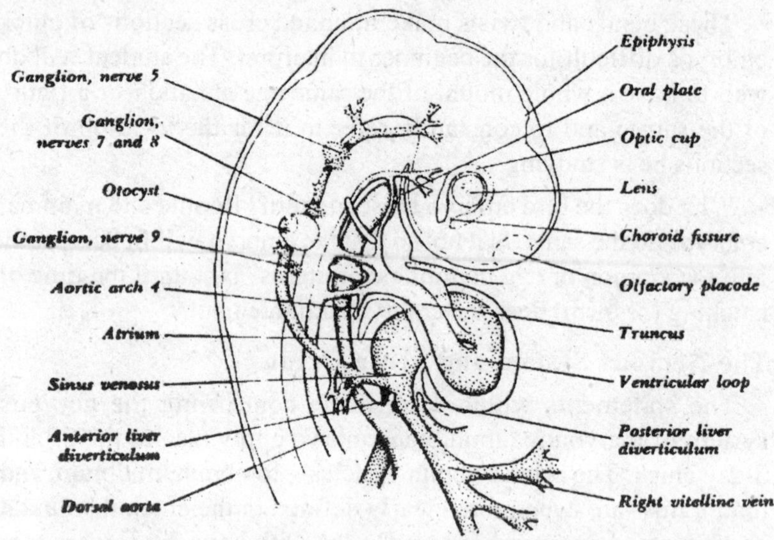

Figure 5.6 : Anterior region of a 48-hour chick from the right side.

The *epiphysis* appears at about the 48th hour, or soon thereafter, as a small upward pocket from the roof of the diencephalon. The *embryonic infundibulum* similarly is a downward pocket from the floor of the diencephalon. It develops in close contact with Rathke's pocket (embryonic hypophysis).

The midbrain (mesencephalon) of the young chick embryo is large and rounded in anticipation of its importance later as the center for reflexes based on sight. The cranial flexure takes place in the region of the midbrain. The result is that transverse sections of 48-hour chicks cut through the midbrain first. Then in sequence the serial sections cut simultaneously anteriorly toward the tip of the forebrain and posteriorly toward the midbrain and spinal cord. The same principle applies to the 72-hour chick except that in this case, because of the cervical bending, the transverse sections first encounter the thin roof of the hindbrain.

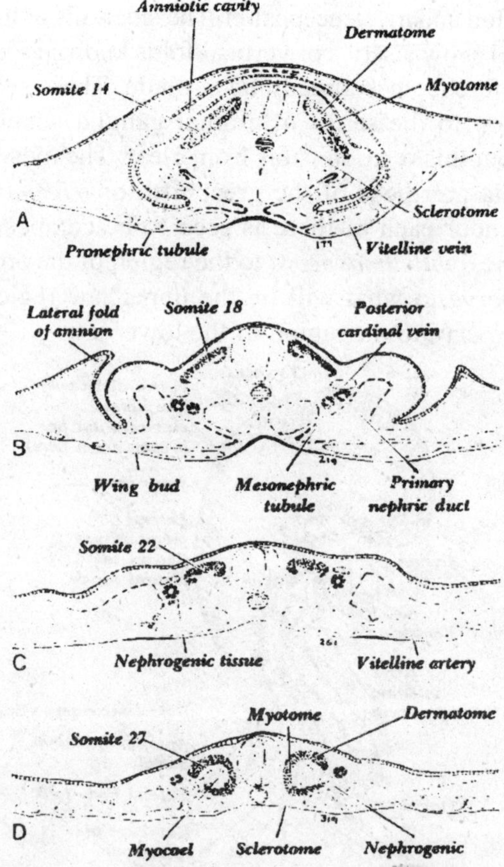

Figure 5.7 : Transverse sections of a 48hour chick. A. Section through pronephric tubules and the 14th somites. 8. Section through the anterior limb buds, mesonephric tubules, and the 18th somites. C. Section through the vitelline arteries and the 22d somites. D. Section through the newly formed 27th somites.

At the 72d hour it is possible to recognize the roots of the third cranial nerves, the *oculomotor nerves,* emerging from the floor of the midbrain.

The *isthmus* is a narrowed region of the brain which marks the division between the midbrain and the hindbrain.

The roof of the hindbrain is thin and shaped like a kite. Its place of greatest width marks the division of the hindbrain into metencephalon and myelencephalon. The sidewalls of the hindbrain are thick and show a series of vertical folds known as neuromeres, to which reference has already been made. These seem to have some relation to the series of cranial ganglia which flank the hindbrain, but this relation is far from clear. The most anterior of these ganglia are those of the great fifth or *trigeminal nerves*. At the 72d hour each of these is seen to be composed of three divisions: the *ophthalmic nerve* to the region of the optic cup, the *maxillary nerve* to what will be the upper jaw region, and the *mandibular nerve* to the region of the lower jaw.

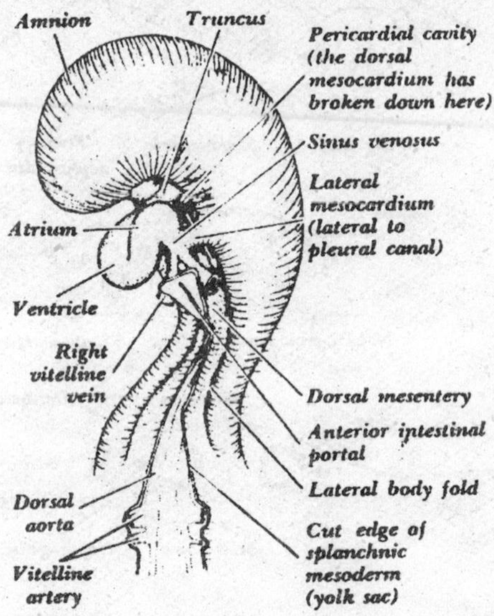

Figure 5.8 : The 48-hour chick within the amnion as seen from the ventral side. The yolk sac has been cut away at the borders of the anterior intestinal portal. The heart is seen protruding into the extra-embryonic coelom. A small white arrow is shown passing through the left pleural canal (medial to the left lateral mesocardium).

Posterior to the ganglia of the fifth nerve are the combined ganglia of the seventh and eighth nerves, the *facial* and *auditory nerves*, respectively. Later in development these ganglia separate; the ganglion of the seventh nerve supplies fibers to the second branchial or hyoid arch, while the eighth ganglion remains closely associated with the *otocyst*.

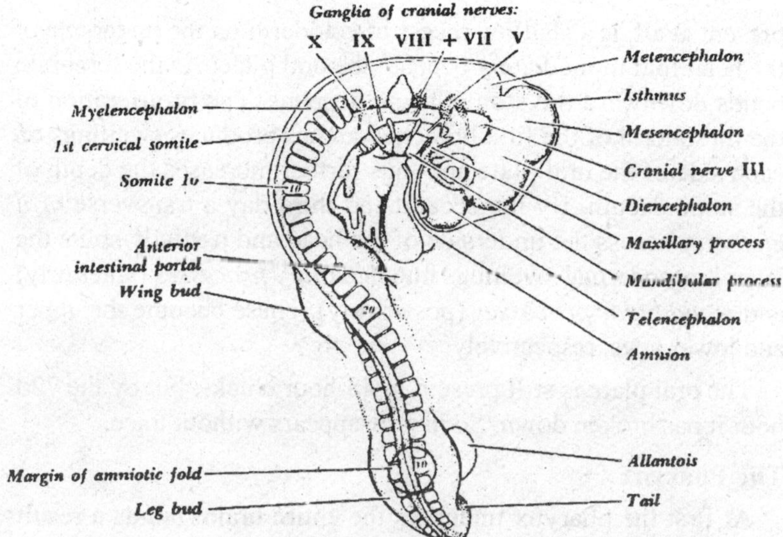

Figure 5.9 : Seventy-two-hour chick embryo from the right side.

At the 48th hour the future otocyst is a pit of epidermis at the side of the hindbrain, open widely to the exterior. But at the 72d hour its opening has closed or almost closed.

A short distance posterior to the otocyst is the ganglion of the ninth or *glossopharyngeal nerve*. It supplies sensory nerve fibers for the third branchial arch. Next comes the ganglion of the tenth or *vagus nerve* whose fibers go to the remaining branchial arches.

The hindbrain narrows gradually to the spinal cord, which is oval in cross section and has thick lateral walls. Its roof plate is thin, and its floor plate is narrow. At the sides of the cord are rows of *spinal ganglia,* arranged at regular intervals like a series of saddlebags. These are not easy to demonstrate at 48 hours, for they are still in the process of formation from the neural crest; but by the 72d hour they are clearly seen lying between each somite and the neural tube.

The posterior neuropore (sinus rhomboidalis) closes (or ceases to exist) about the 40th hour, and from that time on the entire neural tube ends posteriorly at the tail bud.

The Stomodaeum

The stomodaeum of the linear embryo, if one can be said to be present at all, is a shallow recess of ectoderm on the underside of the head fold immediately beneath the oral plate. As the forebrain bends downward the stomodaeum deepens. The proliferation of the mesoderm of the first branchial arches produces swellings on each side of the oral plate and thus further increases the depth of the stomodaeum. By the second and third day a transverse *oral cleft* cuts across the underside of the head and partially splits the lateral mesodermal swellings into *maxillary processes* (anteriorly) and *mandibular processes* (posteriorly). These become the upper and lower jaws, respectively.

The oral plate is still present in 48-hour chicks, but by the 72d hour it has broken down. Soon it disappears without trace.

The Pharynx

At first the pharynx underlies the entire brain; but as a result of the bulging forward of the forebrain and the subsequent formation of the cranial flexure, it comes to underlie the hindbrain only. For most of its length it is broad from side to side, but it tapers to an apex posteriorly. Along each side of the pharynx are the *pharyngeal pouches,* five in number; however, the fifth is rudimentary, little more than an appendage of the fourth. As each pouch pushes laterally through the head mesenchyme it comes into contact with the epidermis of the side of the head. The epidermis responds by forming a *branchial groove* or furrow. Some of the closing plates (gill plates) between the pouches and grooves break through and become open *bronchial clefts* (gill slits). At the 72d hour the first pouch has a dorsal opening, the second pouch has both a dorsal and a ventral opening, while the third pouches have not yet broken through. The fourth pouches never open in the chick.

The *thyroid rudiment* is a downpocketing from the floor of the pharynx, situated approximately midway between the second pair of pharyngeal pouches close to where the ventral aorta from the heart divides into the second and third aortic arches.

The floor of the pharynx posterior to the fourth pouches develops a ventral longitudinal groove, the *laryngotracheal groove.* From the posterior end of this groove the right and left lung buds push out laterally. Gradually, beginning at its posterior end and progressing forward, the groove pinches away from the dorsal part of the pharynx and becomes the *trachea;* but an opening into the pharynx remains at the anterior end, namely, the *glottis*.

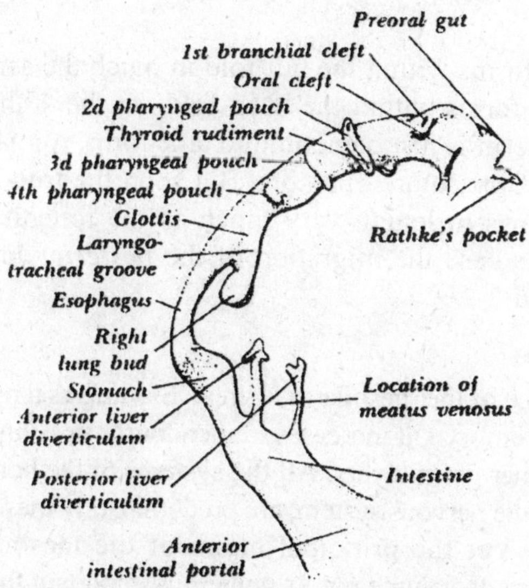

Figure 5.10 : The foregut of a 72-hour chick.

Esophagus, Stomach, and Liver Diverticula

The foregut continues to grow in length by the retreat of the anterior intestinal portal posteriorly. By the 33d hour most of what will be the pharynx has been formed, and the portal is a broad opening from side to side.

By the 48th hour the future *esophagus* and *stomach* have been added, and the portal has narrowed to a vertical slit on the undersurface of the blastoderm.

At this latter stage of development (48th hour), two outpocketings are to be found on the forward wall of the intestinal portal: a dorsal one and a ventral one. These become the *anterior* and *posterior liver diverticula*. The vitelline veins, which run in the lateral folds of the anterior intestinal portal, come together and unite between the diverticula. By the end of the fourth day, the duodenum has been added to the posterior end of the foregut. The liver diverticula open into it. The diverticula, moreover, have given rise to a basket of anastomosing cords which encircle the venous channel.

The Hindgut

The hindgut forms within the tail fold in much the same way that the foregut forms within the head fold. At the 48th hour a pocket of endoderm is just beginning to take form. Actually it is the beginning of the future allantois. But soon the true hindgut appears and grows in length very much as the foregut grows. However, in this case the migration of the *posterior intestinal portal* is forward.

The Mesoderm

The importance of the mesoderm is great. Starting as a relatively small part of the embryo, it increases tremendously in volume until it exceeds the other germ layers. All the systems of the body, with the exception of the nervous system, are predominantly mesodermal in their origin. Yet the principal claim of the mesoderm to preeminence is not its volume nor its numerous important functions, but the fact that it plays the primary role in the establishment of the embryo. It tells the other germ layers what to do.

This is as true of the chick embryo as it is of the amphibian. As the chorda-mesoderm of the primitive streak migrates forward past the primitive node, its regions become specified for their respective fates. When it comes to rest in its final location, it determines the future of the cells of the other germ layers with which it is associated. It is true that the relations between the germ layers are reciprocal, for ectoderm and endoderm are necessary for the normal differentiation of mesoderm. But their roles are different. The mesoderm takes the initiative. The ectoderm and endoderm feed back their influences on the mesoderm.

Head Mesenchyme

The mesenchyme of the head is at first meager, but it increases rapidly, and by the end of the second day of incubation it exceeds the other tissues of the head in volume. It appears to be homogeneous, yet its cells are derived from several embryonic sources: prechordal mesoderm, axial mesoderm anterior to the somites, neural crest, and possibly even the anterior tips of the foregut and notochord. The term mesenchyme, therefore, should be thought of as designating a histological structure rather than a particular embryonic origin. Mesenchyme cells are loose, star-shaped cells capable of migration-although they do not necessarily migrate. In fact they are commonly united into a spongelike mesh in the interstices of which there is a tissue fluid, a sort of thin jelly containing physiological nutrients which serves as a fit medium for cell life.

The Axial Mesoderm (Epimeres)

The axial mesoderm gives rise to somites. At first each somite is a block of columnar cells which radiate out from a central cavity, the *myocoel.* This cavity, however, is more or less filled with loose cells.

The derivatives of each somite are three "-tomes": dermatome, myotome, and sclerotome.

1. The *dermatome* is derived from the lateral wall of each somite, that is, from the wall adjacent to the epidermis. It retains its laminar character for a considerable time, but ultimately (about the fourth day in the case of the anterior somites) it breaks down into mesenchyme and gives rise to the dermis and subcutaneous tissues of the back of the embryo.

2. The dorsal portion of the median wall of each somite constitutes the *myotome.* At first the myotomes are small, but as development proceeds they swing outwardly to a position under the dermatomes. Here they grow, and give rise to the axial muscles of the dorsal side of the body. (The earliest stages are the most posterior.) Their cells, known as myoblasts, elongate longitudinally until they extend the full length of each somite.

3. The ventral portion of the inner wall of each somite, namely, the wall which is adjacent to the notochord and the ventral half of the neural tube, forms the *sclerotome.* As early as the second day of incubation its cells, beginning with the anterior somites, lose their epithelial character and become mesenchyme which migrates

around the neural tube, notochord, and even around the aorta. They multiply rapidly and soon make up a large part of the cross section of the embryo. Their fate is to give rise to the axial skeleton.

The Intermediate Mesoderm (Mesomeres)

The anterior part of the intermediate mesoderm becomes segmented and gives rise to a series of *nephrotomes,* that is, to necks of cells which connect the somites with the lateral mesoderm. In amphibians some of these produce functional *pronephric tubules,* but in the chick the pronephric tubules are vestigial and seemingly never perform as excretory organs. They are to be seen as tiny rudiments, as early as the 33d hour, lateral to somites 8 to 15. Each tubule begins as ar bud which grows outward from the somatic layer of the nephrotome and then hollows out. Its central end retains an opening into the body cavity known as a *nephrostome,* while its lateral end turns backward and unites with the tubule which is next posterior to it. Thus a continuous duct, the *primary nephric duct,* is formed (known also as the pronephric, mesonephric, or Wolffian duct). It grows backward in the space between the somites and the somatic layer of lateral mesoderm. At the 48th hour it has reached the level of about the 25th somite. By the 72d hour it has made contact with the hindgut (cloaca). Shortly afterward it has entered it. The pronephric tubules degenerate, but the primary nephric ducts, to which they give rise, endure.

The intermediate mesoderm posterior to the pronephric tubules does not become segmented into distinct nephrotomes; nevertheless, it is nephrogenic. During the third day of incubation it gives rise to *mesonephric tubules*. These become the functioning excretory tubules of the young embryo. The process can be seen in sections of a 72-hour chick. The most posterior tubules, as usual, are the youngest and consist of spherical condensations of nephrogenous material (cells derived from the mesomere) on the medial side of each primary nephric duct. Farther forward the condensations have become hollow balls of cells. Still farther forward each ball of cells has elongated and become S-shaped, with its lateral end in contact with the primary nephric duct and its medial end encapsuling a tuft of capillaries, the *glomerulus*.

Three features thus distinguish mesonephric tubules from the preceding pronephric tubules: (1) Mesonephric tubules are not segmentally arranged; or rather, several tubules arise from each somite so that segmentation is not apparent. (2) Mesonephric tubules make use of the primary nephric duct which the pronephros produced. (3) Mesonephric tubules begin in capsules which enfold glomeruli, whereas pronephric tubules begin in nephrostomes which open from the body cavity. If there are knots of blood vessels-and they do occasionally occur in the chick-they are in the body cavity external to the tubule.

The Lateral Mesoderm (Hypomeres)

The lateral mesoderm gives rise to the linings of the body cavities-the pericardium, pleura, and peritoneum-but it does far more than this: The somatic layer produces the dermis, muscles, and connective tissues of the lateral and ventral body walls, and also the limb buds. The splanchnic layer gives rise to the epimyocardium of the heart, the walls of the lungs and gut, the cortex of the adrenal gland, and the sex glands. (These matters are discussed further in the chapters which deal with the development of these several organs.)

The Coelom

The development of the coelom is of considerable interest. A ventral view of a 48-hour chick in which the entire splanchnopleure (yolk sac) has been cut away external to the borders of the anterior intestinal portal. This exposes the inner surface of the somatic mesoderm. The lateral body folds are seen pushing ventrally; but as yet they are widely separated, and the heart is shown protruding between them into the extra-embryonic coelom. By the 72d hour the right body fold has pulled all the way across the ventral side of the heart and has fused with the left body fold. As a result, the pericardial cavity is almost completely cut off from the extra-embryonic coelom. The dorsal mesocardium has broken through above the heart, so that the right and left pericardial coeloms communicate with each other.

Lateral to the sinus venosus, the folds of the body wall fuse with the wall of the heart and form bridges, known as the *lateral mesocardia,* between the somatic and splanchnic mesoderms. Across these bridges, the common cardinal veins transport blood from the body wall to the sinus venosus of the heart.

Dorsal and median to each lateral mesocardium, the pericardial and peritoneal coeloms remain in communication with each other by what we shall call the *pleural canals.* Later, the lung buds grow into them, and they become the pleural cavities. Communications between the pericardial and peritoneal coeloms remain open for a brief time, ventral to the lateral mesocardia. But ultimately they close.

Farther posterior, the vitelline veins course within the side folds that bound the anterior intestinal portal. They then enter the sinus venosus. The dorsal and ventral hepatic diverticula, which have outpocketed from the forward wall of the anterior intestinal portal, push between the veins, and together the diverticula and endothelia of the veins give rise to the liver. When the body wall finally becomes complete in this region, it fuses with the underside of the liver and thus forms what is essentially a ventral mesentery. In other words there arises secondarily in the chick a structure (the ventral mesentery) which is primary and original in the lower vertebrates. The lateral mesocardia plus the ventral mesentery are equivalent to the *septum transversum* of mammal embryos.

The Heart and Circulation

During the second and third days of incubation, the area vasculosa continues to expand outwardly across the surface of the yolk. It is bounded at the periphery by the *marginal vein.* Its network of capillaries develops channels which become the vitelline arteries and veins. The arteries diverge to the right and left from the dorsal aorta in the midtrunk region. They transport blood laterally toward the marginal vein and also forward. The blood, having gathered in the forward regions of the area vasculosa, returns posteriorly by way of the *anterior vitelline veins* until it reaches the sidewalls of the anterior intestinal portal. It then turns medially and enters the heart. Later, *lateral vitelline veins* develop, which more or less parallel the vitelline arteries. They lie on the

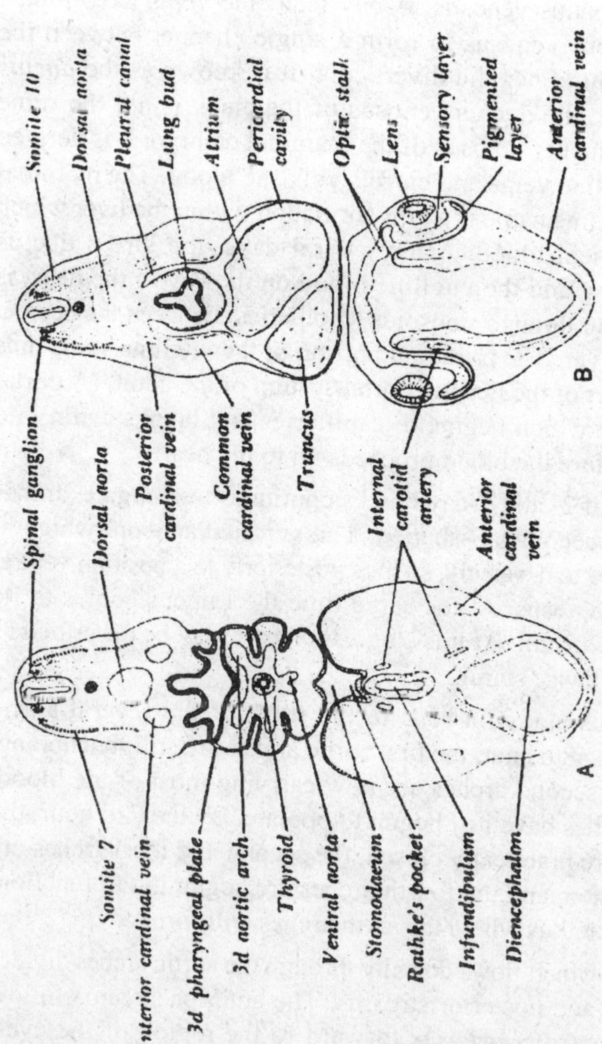

Figure 5.11 : Transverse sections through the 72-hour chick. A. Section through Rathke's pocket and the thyroid rudiment. B. Section through the eyes and lung buds.

mesodermal side of the arteries. The anterior vitelline veins and the marginal veins then gradually disappear.

By the 48th hour of incubation, the anterior intestinal portal has retreated sufficiently for the vitelline veins to unite in the midline and form the sinus venosus. A day later, the veins have joined behind the sinus venosus to form a single channel between the dorsal and ventral hepatic diverticula. It is known as the *ductus venosus*. Now, the ductus venosus of the chick is not the same vessel as the ductus venosus of the mammal embryo. It is derived from the vitelline veins and carries yolk sac blood. The mammal vessel of the same name is a new formation within the liver which carries umbilical blood. After several days, the bird's ductus venosus closes, and the vitelline blood on its way to the heart is forced to detour through sinusoids (capillaries) between the hepatic cords of the liver. The proximal portion of the vitelline veins thus becomes a part of the hepatic portal system of the adult. (A portal system is one which begins in capillaries and breaks again into capillaries before the blood proceeds on to the heart.)

The heart of 2- and 3-day chicks continues to elongate, and as it does so, it becomes S-shaped. The ventricular loop, which at first is oriented transversely, swings posteriorly to a position ventral to the sinus venosus. At the same time the truncus comes to lie ventral to the atrium. At this stage, the heart may be described as a counterclockwise spiral.

The truncus leads forward to the roots of the several aortic arches. At the 48th hour, the first aortic arches are still functioning, although the second arches are now carrying most of the blood. The third arches have just begun to operate. By the 72d hour, the first arches are practically closed, the second and third arches are in full operation, and the fourth arches are beginning to function. Only sprouts tell us where the sixth arches will form.

The blood which flows dorsally through the aortic arches divides into anterior and posterior streams. The anterior streams follow the *internal carotid arteries* forward to the region of the eyes. Here the blood spreads out in a net of capillaries next to the surface of the forebrain. The rest of the blood flows posteriorly in the two

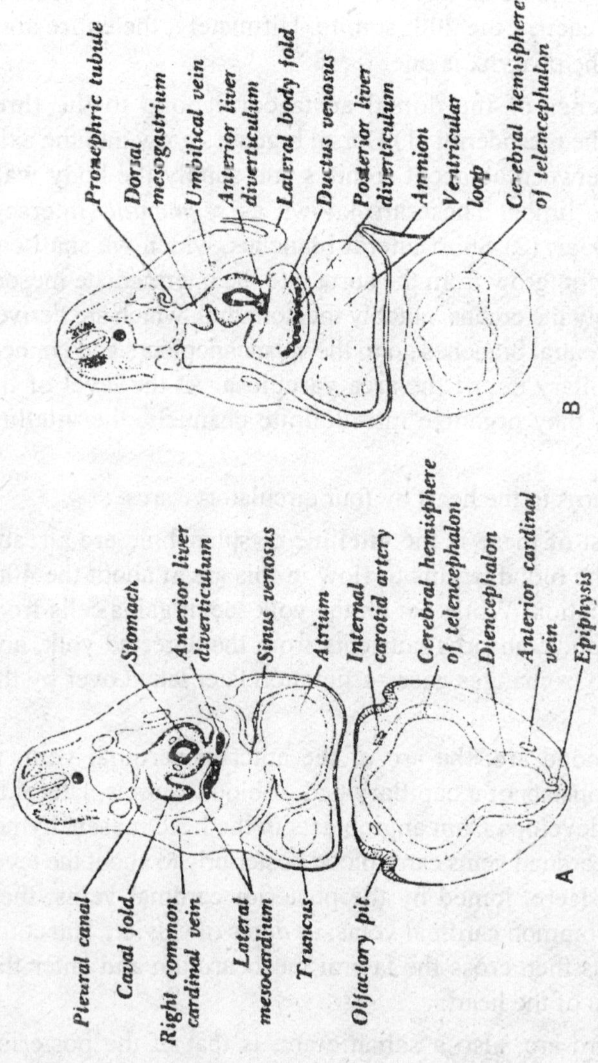

Figure 5.12 : Transverse sections through the 72-hour chick. A. Section through the olfactory pits and lateral mesocardia. B. Section through the liver diverticula and cerebral hemispheres.

dorsal aortas. At the level of the posterior end of the pharynx, the right and left aortas unite in the midline to form a single median dorsal aorta. The process of union progresses posteriorly. At 48 hours, it is complete as far back as the 10th somite. By the 72d hour, it has reached the 20th somite. Ultimately, the entire aorta posterior to the pharynx is one.

The branches of the dorsal aorta correspond to the three divisions of the mesoderm: (1) Dorsal branches grow into the axial mesoderm between adjacent somites and supply the body wall, including the limbs. These are known as *segmental* (intersegmental) *arteries.* (2) Short lateral branches, which we shall call *nephric arteries,* grow from the aorta into the intermediate mesoderm and supply the organs, notably theglomeruli, which are derived from it. (3) Ventral branches go to the splanchnopleure and connect with the capillary net of the area vasculosa. At the level of the 22d somites, they organize into definite channels, the vitelline arteries.

Blood returns to the heart by four circulatory "arcs":

1. The first of these is the vitelline or splanchnic arc already described. The blood begins to flow in this arc at about the 40th hour of incubation. While out on the yolk sac, it gains cells from the blood islands, absorbs nutrients from the digested yolk, and probably also exchanges gases-a function later taken over by the allantois.

2. The second arc, the arc of the anterior cardinal vein, is somatic. It consists of a capillary bed of blood vessels, lateral to the brain. It develops from angioblasts in the head mesenchyme. The anterior cardinal veins carry blood posteriorly to about the level of the heart. Here, joined by the posterior cardinal veins, they become the common cardinal veins, or ducts of Cuvier. The common cardinals then cross the lateral mesocardium and enter the sinus venosus of the heart.

3. The third arc, also a somatic arc, is that of the posterior cardinal veins. At first they bring blood forward from the body wall and tail region. In the trunk, each vein runs just at one side of the intermediate mesoderm and receives blood from the somites

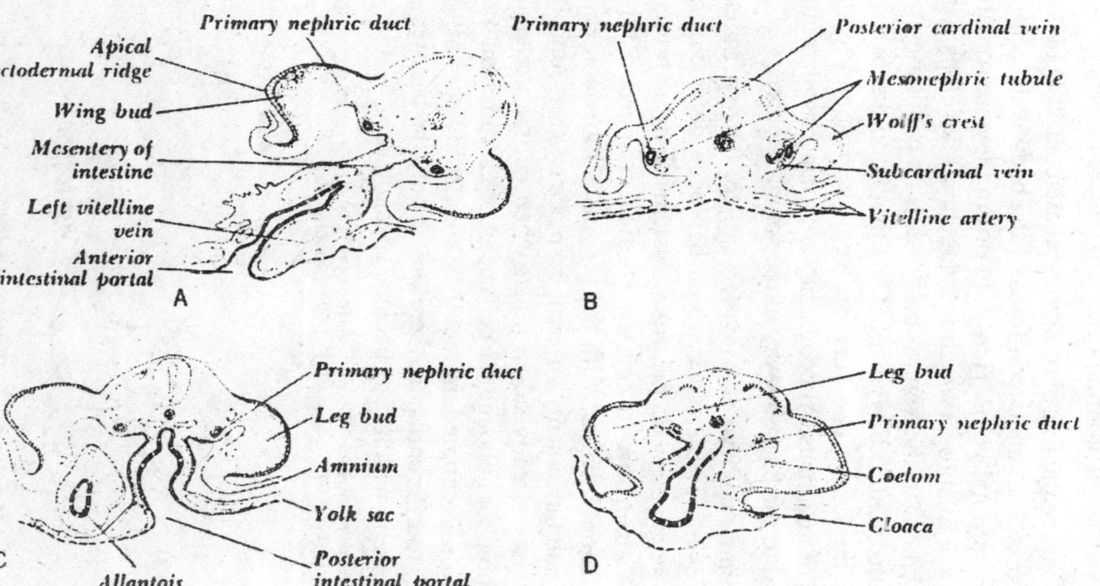

Figure 5.13 : Transverse sections of a 72-hour chick. (These are drawn from a different embryo than the previous sections.) A. Section through the anterior intestinal portal and wing buds. B. Section through the roots of the vitelline arteries. C. Section through the posterior intestinal portal and the leg buds. D. Section through the cloaca.

and nephric tissue. It carries the blood forward to the level of the sinus venosus, where, as we have noted, it joins the anterior cardinal veins and, as the common cardinal, enters the heart.

4. The elements of the fourth or umbilical arc consist at first of a net of blood vessels in the lateral folds of the body wall, ventral to the posterior cardinal veins. Hence, in origin they are somatic vessels. When the allantois (which is splanchnic) develops with its broad attachment to the ventral body wall, its capillaries tap the vessels of the ventral body wall. Thus the blood discovers a direct route to the heart. The allantoic veins correspond to the umbilical veins of mammals. Although they are splanchnic in origin, they go forward to the heart by way of the somatopleure.

The young chick embryo is a favored subject for experimentation. Some results of the work which has been done will be referred to in the chapters on the organ systems. For the present we shall give attention to certain studies concerned with "cytodifferentiation," that is, with the processes by which cells become visibly and functionally different from one another. Actually differentiation has already taken place when cells commence cytodifferentiation, for although the cells of an organ rudiment may appear to bc undifferentiated, in fact they are already committed to their fates. Cytodifferentiation, therefore, is only the last step in the series of steps by which cells become specialized.

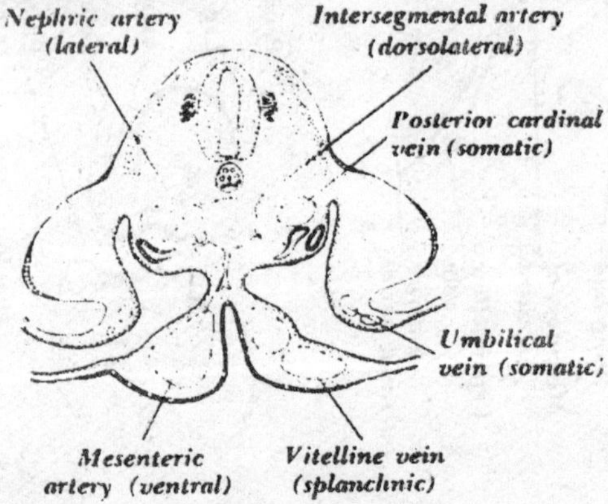

Figure 5.14 : Diagrammatic section through the trunk to illustrate the location of the principal veins and the branches of the aorta.

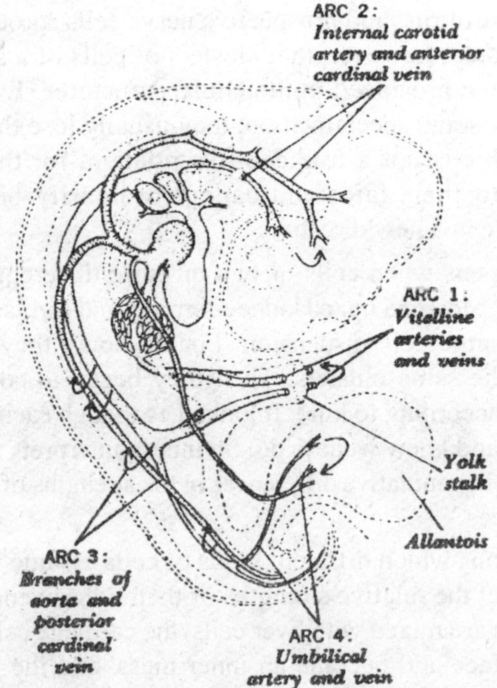

Figure 5.15 : Diagram of the circulation of a 96-hour chick.

In 1952 Moscona treated cells of chick and mouse embryos with calcium-free solutions. He then subjected them briefly to weak solutions containing trypsin or other proteindigesting enzyme. The treatment loosened the bonds between the cells, with the result that they rounded up and became free. After rinsing them in physiological salt solution, he dispersed them in a fluid culture medium containing equal parts of salt solution, chick embryo juice, and chicken serum. The cells settled to the bottom of the culture dish and moved about by means of pseudopod-like processes. When by chance they came into contact one with another, they adhered and drew together. The result was that the bottom of the dish became covered with numerous clusters of squirming cells.

For a day or so the cells which were derived from the various organ rudiments behaved in much the same manner. But then differences appeared. Under suitable culture conditions some of the cells continued to differentiate. Thus future muscle cells produced myofibrils, and prospective nerve cells sprouted axons. It must be noted, however, that clusters of cells of a single kind do not produce organized multicellular structures. Even if they already show some specialization, they usually lose their special character. This is not a true dedifferentiation, for they remain committed to their fates and cannot ordinarily be made to redifferentiate in a new direction.

What happens when cells of two or more different kinds are intermingled? Moscona mixed kidney-forming and cartilage-forming cells in the same cell suspension. For 24 hours they remained mingled in the same clusters. Then they began to sort out and differentiate according to kind. It looked as though each cell knew its own kind and knew what to do. Steinberg interprets the sorting out in terms of quantitative differences in the strengths of intercullar adhesive forces.

The positions which different kinds of cells assume when they sort out reflect the relative strengths of their adhesiveness. When cartilage cells are mixed with liver cells, the cartilage cells migrate from the surface and become an inner mass, and the liver cells remain as an outer coat. When liver cells and cells of the intestinal epithelium are mixed, the liver cells become the inner mass, and the epithelial cells remain on the outside. Thus there is a sort of hierarchy of the strength of adhesion-cartilage > liver > epithelium.

The sorting out not only restores spatial arrangements which often resemble the arrangement of tissues in an embryo, but it results in inductive interactions. As a result, cytodifferentiations take place which do not occur when cells are isolated. We shall refer to several of these interactions in the pages which follow: the interdependence of ectoderm and mesoderm in the production of a limb; the interaction between endoderm and the investing mesenchyme of the gut; the mutual response of epidermis and dermal mesenchyme, by reason of which the skin becomes organized; and the interaction of nephric epithelium and nephric mesenchyme in the production of the finer structure of kidney tissue.

What happens when cells of different species of animals are

mixed together? One might expect that they would separate and each go its own way. This, indeed, is what happens when cells of different species of sponges are intermingled. But Moscona found otherwise when he mixed chick and mouse kidney cells or chick and mouse cartilage cells. Separation according to tissue took place, but segregation according to species did not occur. On the contrary, cells of like type cooperated in the production of "chimeric" structures in which some of the cells were chick cells and some were mouse cells. It is clear that the signals (epigenetic factors) which lead to differentiation are the same in birds and mammals. Wilde obtained similar results when he mixed myoblasts (future muscle cells) of a mouse and chick. The myoblasts fused together to form multinucleate striated muscle fibers. Some of the nuclei in each fiber were chick nuclei, and some were mouse nuclei.

Experiments on the reaggregation of dissociated cells make us aware of activity at two levels of organization: the cellular level and the supracellular, or organismic, level. At the cellular level individual cells are the actors. Yet we must not forget that individual cells acquirk+ their distinctive characters when they were unspecialized parts of a larger whole. Moreover, when cells are isolated they behave as though they are still parts of a whole which no longer is p esent. Indeed, they fulfill their destinies only when they act and interact within the framework of the greater community. The community is as real as the cell, and the language which describes it is as valid. The fact that the organism disappears when it is analyzed does not negate this; so does the cell disappear when it is reduced to its parts.

6

Chick Embryo at Fifty-Five Hours

In chicks which have been incubated from 50 to 55 hours the entire head has beep freed from the yolk by the progress caudad of the subcephalic fold. Torsion has involved the whole anterior half of the embryo and is completed in the cephalic region, so that the head now lies left side down on the yolk. The posterior half of the embryo is still in its original position, ventral prone on the yolk. At the extreme posterior end, the beginning of the caudal fold marks off the tail region of the embryo from the extra-embryonic membranes. The head fold of the amnion has progressed caudad, together with the lateral amniotic folds, impocketing the embryo nearly to the level of the omphalomesenteric arteries.

The *cranial flexure,* which was seen beginning in chicks of about 38 hours, has increased rapidly until at this stage the brain is bent nearly double on itself. The axis of the bending being in the midbrain region, the mesencephalon comes to be the most anteriorly located part of the head, and the prosencephalon and myelencephalon lie opposite each other, ventral surface facing ventral surface. The original anterior end of the prosencephalon is thus brought in close proximity to the heart, and the optic vesicles and the auditory vesicles are brought opposite each other at nearly the same antero-posterior level.

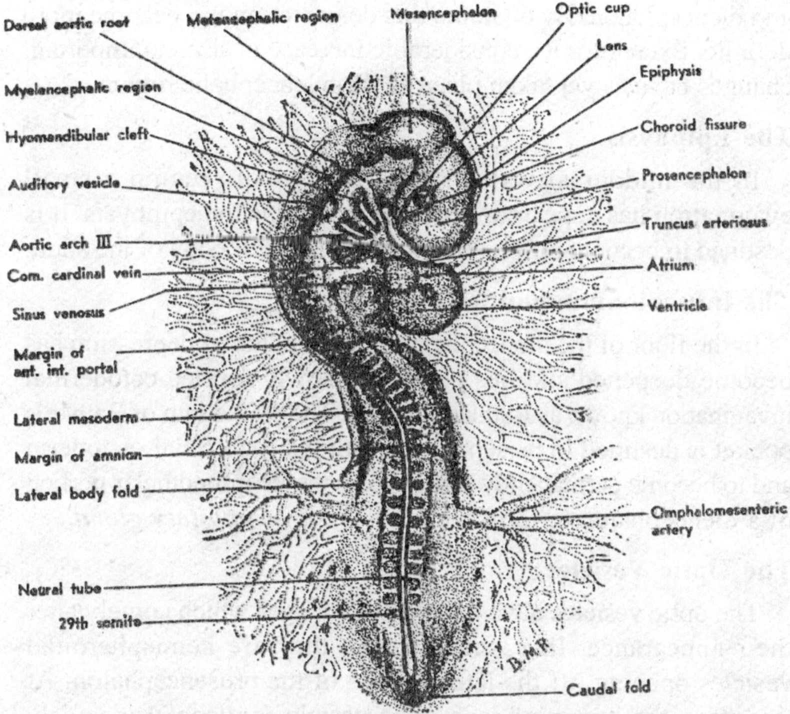

Figure 6.1 : Dextro-dorsa view (X17) of entire embryo of 29 somites (about 55 hours' incubation).

At this stage flexion has involved the body farther caudally as well as in the brain region. It is especially marked at about the level of the heart in the region of transition from myelencephalon to spinal cord. Since this is the future neck region of the embryo, the flexure at this level is known as the cervical flexure.

THE NERVOUS SYSTEM

Growth of the Telencephalic Region

The completion of flexion and torsion in the cephalic region is accompanied by marked changes in the configuration of the brain. The same fundamental regions can, however, be identified throughout this range of development. In embryos of 50 hours a-slight indentation in the dorsal wall of the prosencephalon indicates the impending division of this part of the brain into telencephalon and diencephalon. By 60 hours this demarcation has become more definite. Except for its. considerable increase in size, no important changes have as yet taken place in the telencephalic region.

The Epiphysis

In the middorsal wall of the diencephalic region a small evagination has appeared. This evagination is the epiphysis. It is destined to become differentiated into the pineal body of the adult.

The Infundibulum and Rathke's Pocket

In the floor of the diencephalon the infundibular depression has become deepened and lies close to a newly formed ectodermal invagination known as Rathke's pocket. The epithelium of Rathke's pocket is destined to be separated from the superficial -ectoderm and to become permanently associated with the infundibular portion of a diencephalon to form the *hypophysis* or *pituitary gland.*

The Optic Vesicles

The optic vesicles have undergone changes which comely alter their appearance. In 33-hour chicks they are hemispheroidal vesicles opening off the lateral walls of the prosencephalon. At this stage the lumen of each optic vesicle *(opticoele) is* widely continuous with the lumen of the prosencephalon *(prosocoele)*. The constriction of the optic stalk which begins to be apparent in 38-hour embryos is much more marked in 55-hour chicks.

The most striking and important advance in their development is the invagination of the distal ends of the single-walled optic vesicles to form double-walled *optic cups.* The concavities of the cups are directed laterally. Mesially the cups are continuous, over the narrowed *optic stalks,* with the ventro-lateral walls of the diencephalic region of the original prosencephalon. The invaginated layer of the optic cup is termed the *sensory layer* because it is destined to give rise to the sensory layer of the retina. The layer against which the sensory layer comes to lie after its invagination is termed the *pigment* layer because it gives rise to the pigmented layer of the retina. The double-walled cups formed by invagination are also termed *secondary optic vesicles* in distinction to primary optic vesicles, as they are called before their invagination. The

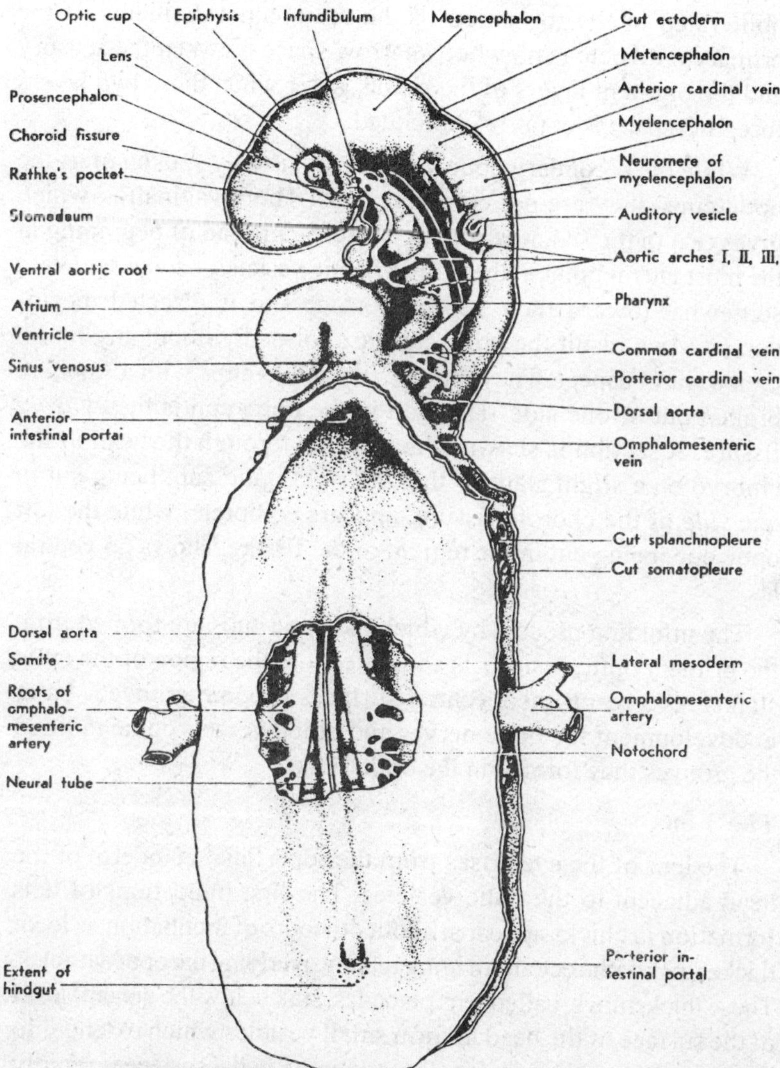

Figure 6.2 : Diagram of diction of chick of about 50 hours. The splanchnopleure of the yolk-sac cephalic to the anterior intestinal portal, the ectoderm of the left side of the head, and the mesoderm in the pericardial region have been dissected away. A window has been cut in the splanchnopleure of the dorsal wall of the midget to show the origin of the omphalomesenteric arteries.

formerly capacious lumen of the primary optic vesicle is practically obliterated in the formation of the optic cup to remains of the primary opticoele is now but a narrow space between the sensory and the pigment layers of the retina. Later when these two layers fuse, this space is entirely obliterated.

While the secondary optic vesicles are usually spoken of as the optic cups, they are not complete cups. The invagination which gives rise to the secondary optic vesicles, instead of beginning at the most lateral point in the primary optic vesicles, begins at a point somewhat toward their ventral surface and is directed mesio-dorsad. As a result the optic cups are formed without any lip on their ventral aspect. They may be likened to cups with a *segment* broken out of one side. This gap in the optic cup is the *choroid* fissure. A section is shown which passes through the head of the embryo on a slight slant so that the right optic cup, being cut to one *side* of the choroid fissure, appears complete; while the left optic cup, being cut in the region, of the fissure, shows no ventral lip.

The infolding process by which the optic cups are formed from the primary optic vesicles is continued into the region of the optic stalks. As a result their ventral surfaces become grooved. Later in development the optic nerves and blood vessels come to lie in the grooves thus formed in the optic stalks.

The Lens

The lens of the eye arises from the superficial ectoderm of the head adjacent to the optic vesicles. The first indications of lens formation in chicks appear at about 40 hours of incubation as local thickenings of the ectoderm immediately overlying the optic vesicles. These thickenings, called lens placodes, sink below the general level of the surface of the head to form small vesicles which extend into the secondary optic vesicles. Their opening to the surface is rapidly constricted, and eventually they are disconnected altogether from the superficial ectoderm. At this stage the opening to the outside still persists, although it is very small. In sections which do not piss directly through the opening, the lens vesicle appears as if it were completely separated from the overlying ectoderm.

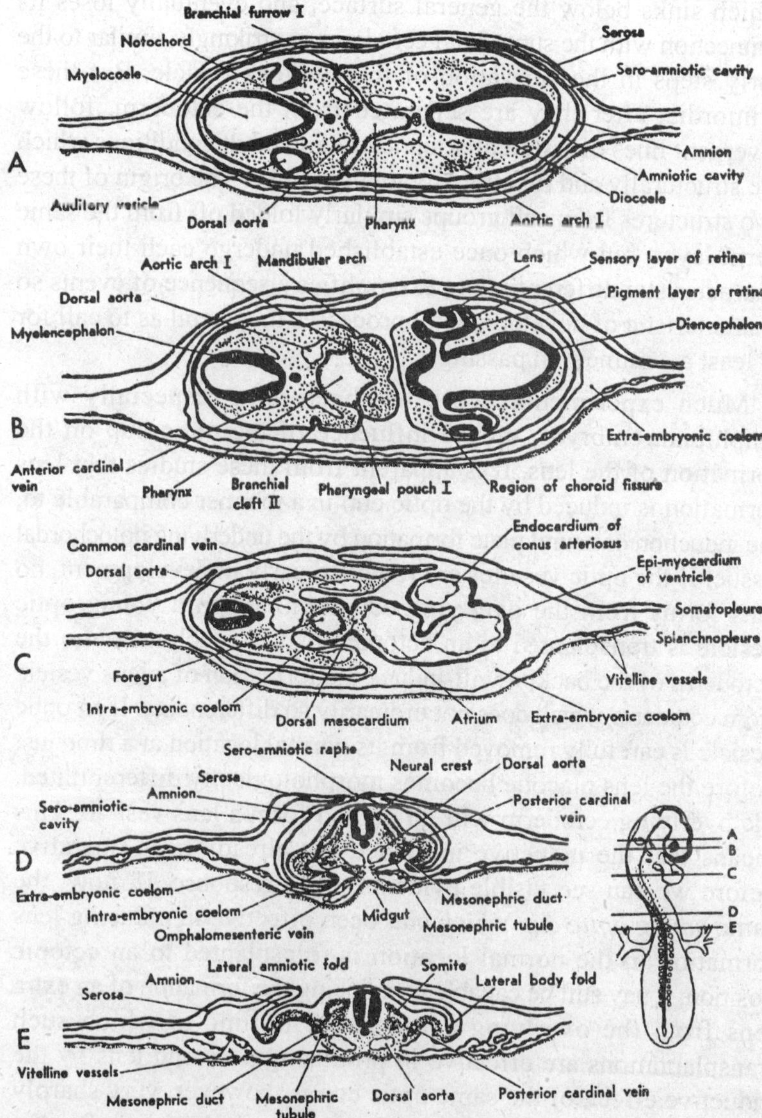

Figure 6.3 : Diagram of transverse sections of 55-hour (30-somite) chick. The locations of the sections are indicated on an outline sketch of the entire embryo.

The derivation of the lens from a placode of thickened epithelium which sinks below the general surface, and eventually loses its connection with the superficial ectoderm, is strikingly similar to the early steps in the derivation of the auditory vesicle. But these primordia, after they are separated from the ectoderm, follow divergent lines of differentiation leading to adult conditions which are structurally and functionally totally unlike. The origin of these two structures from cell groups similarly folded off from the same germ layer, but which once established undergo each their own characteristic differentiation, exemplifies a sequence of events so characteristic of developmental processes in general as to call for at least a comment in passing.

Much experimental work has been done, especially with amphibian embryos, on the influence of the optic cup on the formation of the lens. It is apparent from these studies that lens formation is induced by the optic cup in a manner comparable to, the induction of neural plate formation by the underlying notochordal tissue. If the optic vesicles are removed early in development, no lens forms from the overlying ectoderm. When a young optic vesicle is transplanted to an ectopic position, such as under the ectoderm of the back, it will-induce the formation of a lens vesicle from ectoderm which does not ordinarily so differentiate. If an optic vesicle is carefully removed from its normal location at a time just before the lens placode becomes morphologically differentiated, the overlying ectoderm will go on and form a lens vesicle. This means that the inductive influence has already been operative before we can see visible evidences of a response. If, now, the same *young optic cup which had* been effective in inducing lens formation in the normal location is transplanted to an ectopic position, it nay still be capable of inducing the formation of an extra lens from the overlying ectoderm. The time in which such transplantations are effective in producing a second lens by the inductive effect, of the same optic cup is, however, very sharply limited. If the *optic* cup is moved to its ectopic location after the formation of the normal lens is well under way, there is no response from the ectoderm overlying it in its new position. This means that the inductive effect is operative at only a certain limed period of development. If anything interferes with its *action* at that time it can never again exert its directive influence.

This makes it evident that there *is* a factor of "timing" in developmental processes that is of more far-reaching significance than is generally recognized. Disturbances in the timing of one developmental process with reference to another are undoubtedly involved in the genesis of many types of developmental abnormalities. For an embryo to develop normally it is necessary not only that all the *essential* growing materials be available in the right quantities and at the right places, but also that they must be there at the right time. We, *can* make the crude comparison of a growing embryo with a building contractor, whose plans call for the laying of steel-reinforced concrete at a certain *place* at a certain time. If the cement is mixed, ready to pour, and the truck bringing in the steel reinforcing goes into the ditch in transit, the cement may harden and -become unpourable before the necessary steel arrives.

The Posterior Part of the Brain and the Spinal Cord Region of the Neural Tube

Caudal to the diencephalon the brain shows no great change as compared with the last stages considered. The mesencephalon is somewhat enlarged and the constrictions separating it from the diencephalon cephalically, and the metencephalon caudally, are more sharply marked. The metencephalon is more dearly marked off from the myelencephalon, and its roof is beginning to show thickening. In the myelencephalon the neuromeric constrictions are still evident in the ventral and lateral walls. The dorsal *wall has' becorfie* much thinner than the ventral and lateral walls and shows no trace of division between the neuromeres.

In the *spinal,* cord *region* of the neural tube the lateral walls have become thickened at the expense of the lumen so that the neural canal appears slit-like in sections of *embryos* of this age rather than elliptical as it is immediately after 'the *closure* of the neural folds. At this stage the closure of the neural tube is completed throughout its entire length. The last regions to *close were* at thecephalic and caudal ends of the neural groove. In younger *stage's where they* remained open these regions were known as the anterior *neuropore* and' the sinus rhomboidalis, respectively.

The Neural Crest

In the closure of the neural tube the superficial ectoderm which *at first lay on* ether *side* of the neural groove, continuous with the neural *plate* ectoderm, becomes fused in the midline and separated from the neural plate to constitute an unbroken ectodermal covering. At the same time the lateral margins of the neural plate become fused to complete the *neural* tube. There are cells lying originally at the-edges of the neural folds which are not involved in the fusion of either the superficial ectoderm or the neural plate. These cells form a pair of longitudinal aggregations extending one on either side of the middorsal line in the angles between the superficial ectoderm and the neural tube. With the fusion of the edges of the neural folds to complete the neural. tube, and the fusion of the superficial ectoderm dorsal to the neural tube, these two longitudinal cell masses become for a time confluent in the midline . But it should be emphasized that this aggregation of cells arises from paired components and soon again separates into. right and left parts. On account of its temporary position dorsal to the neural tube it is known as the *neural crest.*

The neural crest-should not be contused with the margin of the neural fold with which it is *associated* before the closure of the neural tube. The margin of the neural fold involves cells which go into the superficial ectoderm and into the neural tube, as well as those which are concerned in the formation of the neural crest.

When first established, the neural crest is continuous antero-posteriorly. A development proceeds, the cells of the neural crest migrate ventro laterally on either side of the spinal cord and at the same time become segmentally clustered. The segmentally arranged cell groups thus derived from the neural crest: give rise to the dorsal root ganglia of the spinal nerves, and in the cephalic region to the ganglia of the sensory cranial nerves.

THE DIGESTIVE TRACT

The Foregut

The, manner in which the three primary regions of the gut tract.

are established has already been considered in a general way. In 50- to 55-hour chicks the foregut has acquired considerable length. It extends, from the anterior intestinal portal cephalad almost to the region of the infundibulum.

As the first part of the tract to be established, the foregut is naturally the most advanced in differentiation. We can already recognize a pharyngeal and an esophageal portion. The pharyngeal region lies ventral to the myencephalon and is encircled by the aortic arches. The pharynx is somewhat flattened dorso-ventrally and has a considerably larger lumen than the esophageal part of the foregut.

The Stomodaeum

There is at this stage no mouth opening into the pharynx. However, the location where the opening will be formed is indicated by the approximation of a ventral outpocketing near the anterior end of the pharynx to a depression farmed in the adjacent ectoderm of the ventral surface of the head. The ectodermal depression, known as the stomodaeum, deepens until. its floor lies in contact with the entoderm of the pharyngeal outpocketing. The thin layer of tissue formed by the apposition of the stomodaeal ectoderm to the pharyngeal entoderm is known as the oral plate. Later in development the oral plate breaks through bringing the stomadaeum and the pharynx into open communication and thereby establishing the oral opening.

The Pre-oral Gut

The oral opening is not established at the extreme cephalic end of the pharynx. The part of the pharynx which extends cephalic to the mouth opening is known as the *pre-oral gut.* After the rupture of the oral plate, the pre-oral gut eventually disappears, but an indication it persists for a time as a small diverticulum termed Seessel's pocket

The Midgut

Although the midgut is still the most extensive of the three primary divisions of the digestive tract, it presents little of interest. It is nothing more than a region where the gut still lies open to the yolk. It does not have even *a* fixed identity. As fast as any part of the midgut acquires a ventral wall by the closing-in process involved in the progress of the subcephalic *and* subcaudal folds, it ceases to be midgut and becomes foregut or lundgut. Differentiation and local specializations appear in the digestive tract only in regions which have ceased to be midgut.

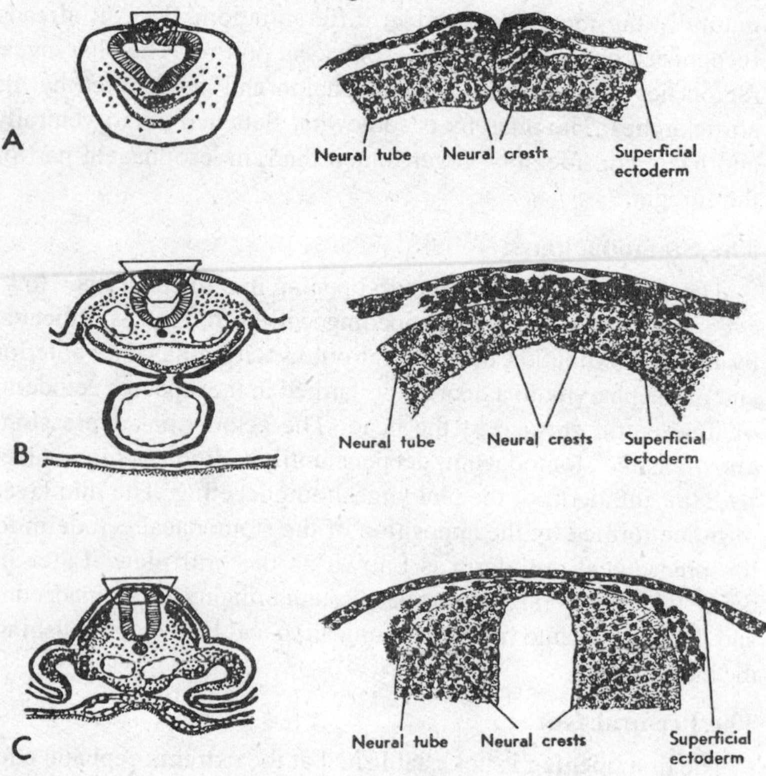

Figure 6.4 : Drawings from transverse sections to show the origin of neural crest cells. The The location of the area drawn is indicated on the small sketch to the left of each drawing. (A) ;Anterior rhombencephalic region of 30-hour chick. (B) Posterior rhombencephalic region of 36-hour dock. (C) Middorsal region of cord in 55-hour chick.

The Hindgut

The hindgut first appears in embryos of about 50 hours. The method of its formation is similar to that by which the foregut was established. The subcaudal fold undercuts the tail region and walls off a gut pocket the entrance into which is the *posterior intestinal portal*. The hindgut is lengthened at the expense of the midgut as the subcaudal fold progresses cephalad and is also lengthened by its own growth caudal. It-shows no local specializations until later in development.

At this stage the chick embryo has unmistakable branchial (gill) arches and branchial (gill) lefts. Although only transitory, they are morphologically of great importance not only from the comparative viewpoint and because of their significance as structures exemplifying recapitulation, but also because of their involvement in the formation of the embryonic arterial system, some of the ductless glands, the Eustachian tube, and the face and jaws.

The branchial clefts are formed by the meeting of ectodermal depressions, the branchial furrows, with diverticula from the lateral walls of the pharynx, the pharyngeal pouches. During most of the time the branchial furrows are conspicuous features in entire embryos, they may be seen, by studying sue, to be closed by a thin layer of tissue composed of the ectoderm of the floor of the branchial furrow and the entoderm at the distal extremity of the pharyngeal pouch. The breaking through of this thin double layer of tissue brings the pharyngeal pouches into communication with the branchial furrows, thereby establishing open branchial clefts. In birds-an open condition cf the clefts is transitory. In the chick the most posterior of the series of clefts never becomes open. Although some of the clefts never become open and others open for a short time, the term cleft is usually used to designate these structures which, whether open or not, are clearly homologous with the gill clefts of water-living ancestral forms.

The position of the branchial or gill clefts is best seen in entire embryos. They are commonly designated by number beginning with the first deft posterior to the mouth and proceeding caudad. The first post-oral cleft appears earliest in development and is discernible at about 46 hours of incubation. Branchial deft II,appears soon after, and by 50 to 55 hours three clefts have been formed.

Between adjacent branchial clefts, the lateral body-wads about the pharynx are thickened. Each of these lateral thickenings meets and merges in the midventral line with the corresponding thickening of the opposite side of the body. Thus the pharynx is encompassed laterally and ventrally by a series of arch-like thickenings, the branchial, or gill, arches. The branchial arches, like the branchial clefts, are designated by number, beginning at the anterior end of the series. Branchial arch I lies cephalic to the first post-oral deft, between, it and the mouth region. Because of the part it plays in the formation of the mandible, it is also designated as the *mandibular arch.* Branchial *arch II* is frequently termed the *hyoid arch*, and visceral cleft I, because of *its position* between the mandibular and hyoid arches, is known as the *hyomandibular cleft.* Posterior to the hyoid arch the branchial arches and clefts are ordinarily designated by their post-oral numbers only.

These are other structures which are just beginning to be differentiated in the pharyngeal, region and foregut of embryos of this stage, but it seems better to consider them in connection with later stages when their significance will be more readily grasped.

THE CIRCULATORY SYSTEM

The Heart In, embryos of 30 to 40 hours incubation we traced the expansion of the heart till it was bent to the right of the embryo in the form of a U-shaped tube . The disappearance of the dorsal mesocardium, except at its posterior end leaves the midregion of the heart lying unattached and extending to› the right, into the pericardial region of the coelom. The heart is fixed with reference to the body of the embryo at its cephalic end, where the ventral aortic roots lie embedded beneath the floor of the pharynx, and caudally *in* the sinoatrial region, where it is attached by the omphalomesenteric veins, the common cardinal veins, and the per*slstent* portion of the dorsal mesocardium.

During the period between 30 and 55 hours of incubation the heart itself is growing more rapidly than is the body of the embryo in the region where the heart lies. Since its cephalic and caudal ends are fixed, the unattached midregion of the heart must undergo bending. It becomes at first U-shaped and then twistedon itself to form a loop. The atrial region of the *heart* is *forced* somewhat to the, left, and the truncus is thrown across the atrium by being twisted to the right and dorsally. The ventricular region constitutes the loop proper 74). This twisting process reverses the original cephalo-caudal relations of the atrial and ventricular regions. Before the twisting, the atrial region of the heart was caudal to the ventricular region as it is *in the* adult fish heart. In the twisting of the heart the atrial region, by reason of its association with the fixed sinus region of the heart, undergoes relatively little change- in position. The ventricular riots is carried over the dextral side of the atrium and comes to lie caudal it, thus arriving in the relative position it occupies in the adult heart of birds and mammals.

The bending and subsequent twisting of the heart lead toward its division into separate chambers. As yet, however, no indication of the actual partitioning f the heart into right and left sides is apparent. It is still essentially a tubular *organ* through which the blood passes directly, without any division into separate channels.

The Aortic Arches

In 33- to 38-hour chicks the ventral aortae communicate *wit the* dorsal aortae over a single pair of aortic arches which bend around the anterior end of the pharynx. With the formation of the brachial: arches new aortic arches appear. The original pair of aortic arches comes to lie in the mandibular arch. New aortic arches are formed caudal to the first pair, one pair in each branchial arch. In chicks of 50 hours, two pairs of aortic arches have been established, and *frequently* the third also is present or starting to form. Usually by the fourth aortic arch is beginning to take shape.

The Fusion Pool of the Dorsal Aortae

The dorsal aortae arise as vessels paired throughout their entire length. As development progresses they fuse in the midline to form the unpaired dorsal aorta familiar in adult anatomy. This fusion takes place a first at cardiac level. Cephalically it never extends to the pharyngeal region. Caudally the whole length of the aorta is eventually involved. By 55 hours of incubation the fusion has progressed caudad to about the level: of the fourteenth somite.

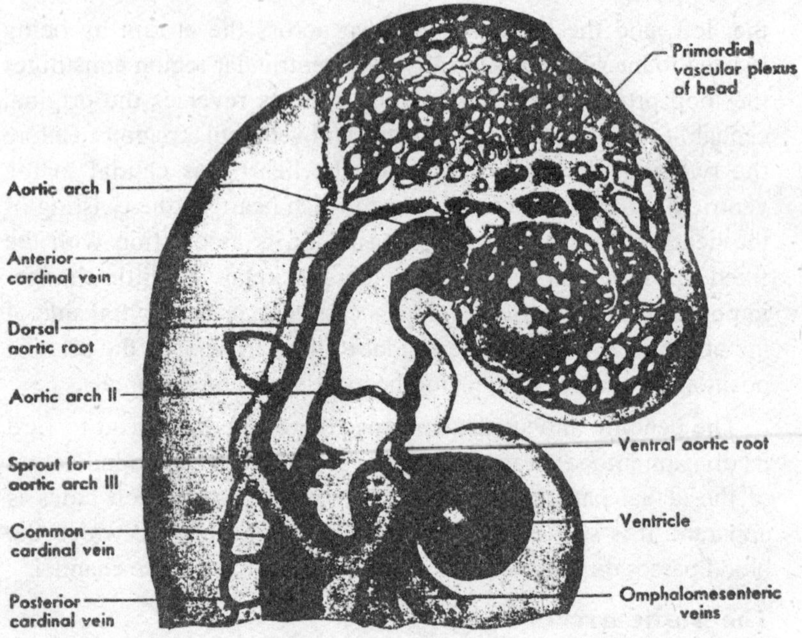

Figure 6.5 : Dextral view of cephalic and cardiac regions of injected chick of about 50 hours' incubation. This figure shows a later stage in the development of the anterior cardinal vein from the prima p capillary plexus of the cephalic regioh. The main channel, only vaguely suggested in the previous figure is hem quite definite. Here also a second aortic arch has been completed, and a plexiform outgrowth of vascular endothelium from the dorsal aortic root toward the ventral aorta indicates the impending formation of the third aortic arch.

The Cardinal and Omphalomesenteric Vessels

The relationships of the cardinal veins and the omphalomesenteric vessels are little changed from the conditions seen in 40- to 50-bout chicks. The posterior cardinals have elongated, keeping pace with the cal progress of differentiation in the mesoderm. They lie just dorsal to the intermediate mesoderm in the angle formed between it and the somites. As the omphalomesenteric veins approach the sinoatrial part of the heart, they come to lie progressively closer to each other in the midline. It is at this region of the convergence of the omphalomesenteric vein that the common cardinal veins enter them. dorso-laterally. This region of confluence of the great veins, as we shall see in considering older stages, is being involved in the moulding of the sinus venosus. The omphalomesenteric arteries, meanwhile, are tending to lose the multiple roots by which they emerged from the dorsal aortae in younger stages, but otherwise they exhibit essentially the same relationships.

THE DIFFERENTIATION OF THE SOMITES

When the somites are first formed, they consist of nearly solid masses of cells derived from the dorsal mesoderm. The cells composing them show *a* more *or* less radial arrangement. In the center of the somite a cavity is usually discernible. *This* cavity is at first extremely minute, and in somites which have been recently formed it may be altogether wanting.

As a somite becomes more sharply marked off, the radial arrangement of the outer zone of its cells becomes more definite. The boundaries of the central cavity are considerably extended, but its lumen is almost completely filled by a core of irregularly arranged cells. In sections which pass through the middle of the somite, this central core of cells is seen to arise from the lateral wall of the somite where it is continuous with the intermediate mesoderm.

A little later in development the outer zone of cells on the ventro-mesial face of the somite loses its originally definite boundaries and becomes merged with the central core of cells. This ill-defined cell aggregation, known as the *sclerotome, becomes* mesenchymal : in characteristics, and extends ventro-mesiad from the somite of either side toward the notochord. The cells of the sclerotomes of either side continue to converge about the notochord. Later in development they will take part in the formation of the axial skeleton. ,

During the emergence of sclerotomal cells, the dorso-lateral part of the original outer cell-zone of the somite has maintained its definite boundaries and epithelioid characteristics. The part of this outer zone which thus lies parallel to the ectoderm is known as the *dermatome*. It received this name because. its cells were believed to migrate out and come to underlie the ectoderm, giving rise to the connective tissue layer (dermis) of the integument. Although some cells from this region of the somite undoubtedly

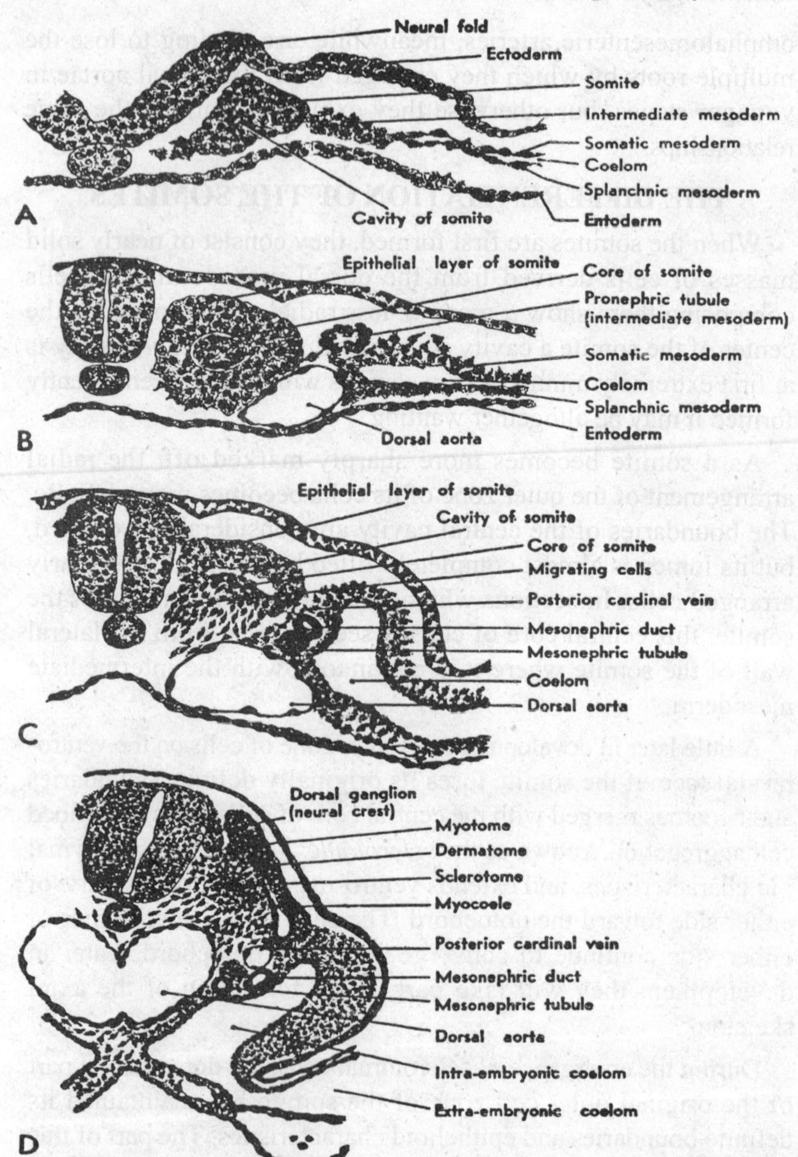

Figure 6.6 : Drawings from transverse sections to show the differentiation of the somites. (A) Second somite of a 4-somite chick; (B) ninth somite of a 12-somite chick; (C) twentieth somite of a 30-somite chick; (D) seventeenth somite of a 33-somite chick.

are contributed to the formation of the deep layers of the skin, the conviction has been gaining ground that many; perhaps most, of them take part in the formation of muscle. Furthermore, the connective tissue layer of the skin is known to receive many cells from the somatic mesoderm generally, and from the diffuse rnesenchyme in the cephalic region where there are no somites. The term dermatome is so firmly fixed that it is probably unwise to attempt to discard it, but we should bear in mind that although its does contribute to the dermis, it probably does not do so any more extensively than other regions of the mesoderm which lie in close proximity to the ectoderm.

The dorsomesial portion of, the outer zone of the somite becomes the myotosne. It is folded somewhat laterad from its original position next to the neural tube and comes to lie ventro-mesial to the dermatome and parallel to it. (A later stage in the differentiation of the somite) The portion of the original cavity which persists for a time between the dermatome and myotonte is termed the *myocoele.* The myotome undergo the most extensive growth of any of the parts of the somite, giving rise eventually to the major part of the skeletal musculature of the body.

The Urinary System

Certain parts of the urinary system which have been established in chcks of 50 to 55 hours will be found located and labeled. The urinary system is relatively late in becoming differentiated and only a few of the early steps in its formation can at this time be made out. Many structures which later become of great importance are not represented even by primoridial cell aggregations. Except for those who are well grounded in comparateive anatomy, any logical discussion of the structures which have appeared mjust anticipate much that occurs later in development. Consideration of the mode of origin and significance of the nephric organs appearing at this stage has, therefore, beed deferred.

7

Chick Embryo on Third and Fourth Day

EXTERNAL FEATURES

Torsion

Chicks of 3 days' incubation have been affected by torsion throughout their entire length. Torsion is complete well posterior to the level of the heart, but the caudal portion of the embryo is not yet completely turned on its side. In 4-day chicks the entire body has been turned through 90 degrees, and the embryo lies with its left side on the yolk.

Flexion

The cranial and cervical flexures which appeared in embryos during the second day have increased so that in 3-day and 4-day chicks the long axis of the embryo shows nearly right-angled bends in the midbrain and in the cervical region. The midbody region of 3-day chicks is slightly concave dorsally. This is due to the fact that the embryo is still broadly attached to the yolk in that region. By the end of the fourth day the body folds have undercut the embryo, so it remains attached to the yolk only by a slender stalk. The yolk-stalk soon becomes elongated, allowing the embryo to become first straight in the middorsal region and then convex dorsally. At the same time the caudal flexure is becoming more pronounced. The progressive increase in the cranial, cervical, dorsal, and caudal flexures results in the bending of the embryo on itself so that its originally straight long axis becomes C-shaped, and its head and tail lie close together.

The Branchial Arches and Clefts In 3- and 4-day embryos a fourth branchial cleft has appeared caudal to the three that were already formed in 55-hour chicks. The branchial arches are thicker and more conspicuous than in younger embryos. In lightly stained whole-mounts of a 3-day chick it is still possible to make out the aortic arches within the branchial arches. In a chick of 4 days the branchial arches have become so much thickened that it is very difficult, by the study of cleared whole-mounts alone, to **see** the vessels traversing them. At this stage the aortic arches must be followed in serial sections in order to get a clear picture of their relations.

The Oral Region

As a result of the cranial and cervical flexures, the pharyngeal region and the ventral surface of the head lie so closely together that it is difficult to make out the topography of the oral region by study of entire embryos. If the head and pharyngeal region are cut from the trunk and- viewed from the ventral aspect, the relations of the structures about the mouth are well shown. The mandibular arch forms the caudal boundary of the oral depression. Arising on either side in connection with the lateral part of the mandibular arch are paired elevations, the maxillary processes, which grow mesiad and form the cephalolateral boundaries of the mouth opening. The *nasal pits* appear as shallow depressions in the ectoderm of the rostral part of the head which overhangs the mouth region. Surrounding each nasal pit is a U-shaped elevation with its limbs directed toward the oral cavity. The lateral limb of the elevation is the *naso-lateral process,* and the median limb is the *naso-medial process.* As development proceeds, the two naso-medial processes grow toward the mouth and meet the maxillary processes which are growing in from either side. The fusion of the two naso-medial processes with each other in the midline and the fusion f each of them laterally with the maxillary process of it own side gives rise the upper jaw *(maxilla).* The fusion in the midline of the right and left components of the mandibular arch gives rise to the lower jaw *(mandible).*

The Allantois

The development of the extra-embryonic membranes has already been considered and needs no further discussion [ere. In order to show the embryos more clearly, the extrabembryonic membranes, except for the allantois, have been removed from the specimens. The cut edge of the amnion shows at its anterior attachment to the body, opposite the anterior appendage-bud and just caudal o the tip of the ventricle. The allantois in the 3-day chick is as yet small and is concealed by the posterior appendage-buds. In 4-day embryos it has indergone rapid enlargement and projects from the umbilical region as a talked vesicle of considerable size.

The Appendage-buds

Both the anterior and posterior appendage-buds ave been well established in embryos of 3 days. The anterior (wing) bud is formed opposite somites 15 to 20 and the posterior (leg) bud develops at the level of somites 27 to 32. By the end of the fourth day the appendage-buds have increased considerably in size to form paddle-shaped extensions from the sides of the body. The main mass of the buds is composed of closely packed mesenchymal cells; the outer covering is of ectoderm. When **seen** in sections, there is a conspicuously thickened band of this ectodermal covering along the convex outer margin of the buds. This is the *apical ectodermal ridge*. It is nearly half through the incubation period before the appendages take on avian characteristics.

There are exceedingly interesting interactions between the mesodermal core and the ectodermal apical ridge of a developing appendage-bud. The earliest indication of impending limb formation is to be seen in chicks around the middle of the third day of incubation. There is then a heightened rate of cell proliferation in the somatic layer of the lateral mesoderm at the levels where appendage-buds are destined to appear. In these regions cells emerge from the parent layer and become mesenchymal in their characteristics. It is these mesenchymal cells which, migrating laterad, constitute the core of the appendage-bud. Moreover they induce the formation of the apical ridge in the overlying ectoderm. We have already considered various examples of such inductive

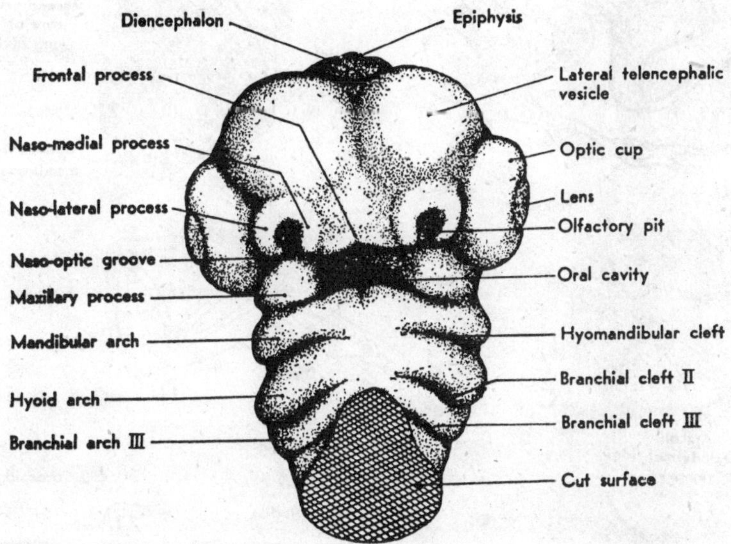

Figure 7.1 : Drawing to show the external appearance of the structure in the oral region of a 4-day chick. Ventral aspect.

activity in other locations. Of particular interest in this instance is the way in which, once established, the ectodermal apical ridge exercises a return controlling action on the mesenchymal core of the bud. If the apical ridge is removed, the distal part of the appendicular skeleton fails to develop. Wing bones fail to form in the anterbior buds or foot bones in the posterior bud. The earlier in development the apical ridge is removed the greater is the difficiency in the development of the distal parts of the appendage. It should be emphasized that such disturbances are limited to the more distal portions of the limbs and that the appendicular girdles are not affected.

Another interesting feature of the development of the appendages is the way in which some of their molding is brought about by selective cell death. By using such vital dyes as Nile blue

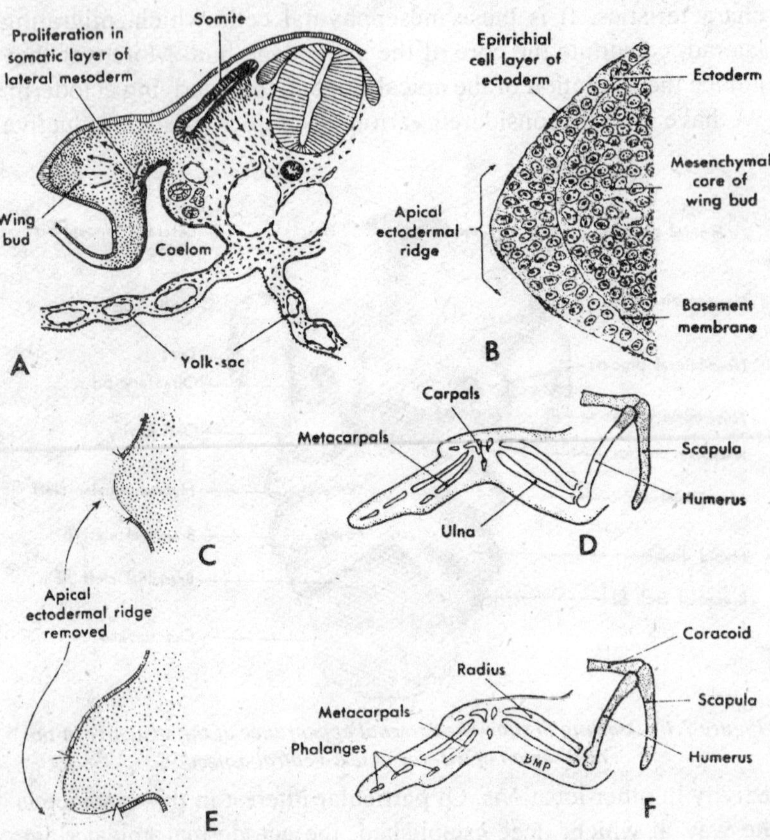

Figure 7.2 : The effect of the removal of the apical ectodermal ridge on the development of the wing-bud. (A) Diagram showing location of the center of mesodermal proliferation involved in forming the core of the wing-bud. (B) Cell detail drawing of tip of wing-bud of 3-day chick to show the apical ridge. (C) Ridge removal in 3-day chick. (D) Skeletal deficiency resulting from third day removal of apical ridge. (E) Ridge removal at 4 days. (F) Skeletal deficiences resulting from fourth day removal of apical ridge. In D and F the parts of the skeleton which develop are stippled; the parts failing to develop are shown in outline only.

sulfate or neutral red, Saunders and his coworkers (1992) have shown areas of cell deterioration to be extensively involved in the shaping of digits or wing parts in the original paddle-shaped appendage-buds. We have been accustomed to thinking of developmental processes almost exclusively in terms of growth activities. The concept of selective cell death as a molding process adds a new dimension to our thinking in regard to morphogenetic processes.

THE NERVOUS SYSTEM

Summary of Development Prior to the Third Day

The earliest indication of the formation of the central nervous system appears in chicks of 16 to 18 hours as a local thickening of the ectoderm which forms the neural plate. The neural plate then becomes longitudinally folded to form the neural groove. By fusion of the margins of the neural folds, first in the cephalic region and later caudally, the neural groove is closed to form a tube and at the same time separated from the body ectoderm. The cephalic portion of the neural tube becomes dilated to form the brain, and the remainder of the neural tube gives rise to the spinal cord.

In its early stages the brain shows a series of enlargements in its ventral and lateral walls, indicative of its fundamental metameric structure. In the establishment of the three-vesicle condition of the brain, the lines of demarcation between prosencephalon, mesencephalon, and rhombencephalon are formed by the exaggeration of certain of the interbneuromeric constrictions and the obliteration of others. The original neuromeric enlargements persist longest in the rhombencephalon.

The three-vesicle condition of the brain is transitory. By 40 hours the division of the rhombencephalon into metencephalon and myelencephalon is clearly indicated. The division of the prosencephalon and the establishment of the five-vesicle condition characteristic of the adult brain do not take place until somewhat later.

In chicks of 55 hours the development of the cranial flexure has resulted in the bending of the brain so that the entire prosencephalon is displaced first ventrad and then caudad toward the heart. At the same time the head of the embryo has undergone torsion and lies with its left side on the yolk. Although flexion and torsion have thus completely changed the general relationships of the brain as seen in entire embryos, the regions already established in 40-hour chicks are still evident. The prosencephalon has, however, become very noticeably enlarged cephalic to the optic vesicles, and a slight constriction in its dorsal wall indicates the beginning of the demarcation of the telencephalic region from the diencephalic region.

Formation of the Telencephalic Vesicles By the end of the third day the antero-lateral walls of the primary forebrain have been evaginated to form a pair of vesicles lying one on either side of the midline. These lateral evaginations are known as the *telencephalic vesicles*. In chicks of 4 days the telencephalic vesicles have undergone considerable enlargement. The openings through which their cavities are continuous with the lumen of the median portion of the brain are later known as the *foramina of Monro*. The telencephalic division of the brain includes not only these two lateral vesicles but also the median portion of the brain from which they arise. The lumen of the telencephalon has therefore three divisions, *a median telocoele,* broadly confluent posteriorly with the diocoele, and two lateral telencephalic vesicles, connecting with the median telocoele through the foramina of Monro.

Before the formation of the telencephalic vesicles the most anterior part of the brain lay in the midline, but the rapid growth of the telencephalic vesicles soon carries them rostrally beyond the median portion of the telocoele. The median anterior wall of the telocoele which formerly was the most rostral part of the brain, and which remains the most rostral part of the brain lying in the midline, is known as the *lamina terminalis*. The telencephalic vesicles become the cerebral hemispheres, and their cavities become the paired lateral ventricles of the adult brain. The hemispheres undergo enormous enlargement in their later development and extend dorsally and posteriorly as well as rostrally, eventually covering the entire diencephalon and mesencephalon under their posterior lobes. They contain the brain-centers for memory and for all actions conditioned by past experience.

As a matter of convenience in dealing with the morphology of the brain, more or less arbitrary lines of division between the

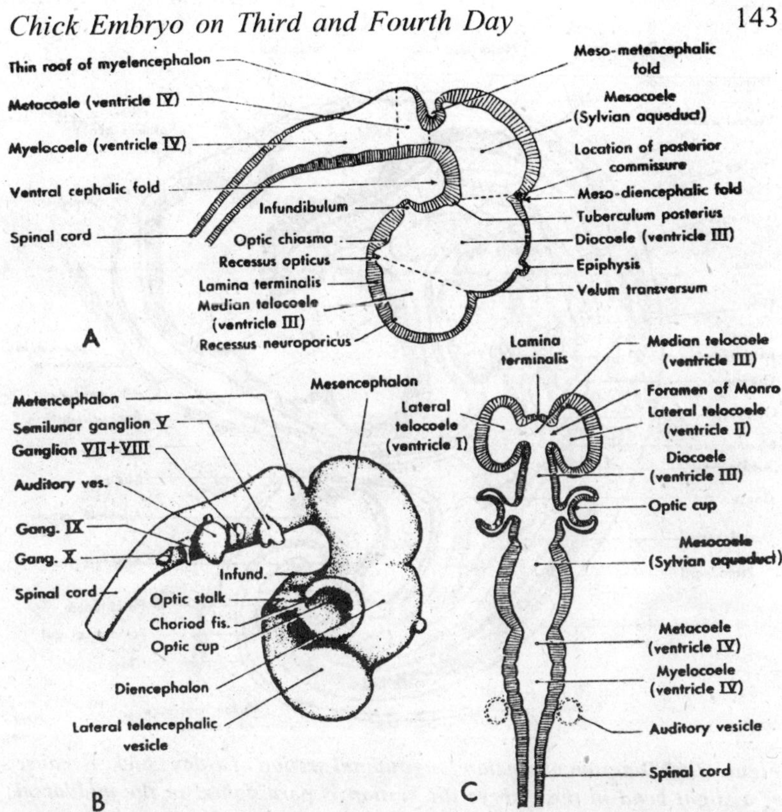

Figure 7.3 : Diagrams to show the topography of the brain of a 4-day chick. (A) Plan of sagittal section. The arbitrary boundaries between the various brain vesicles (according to von Kupffer) are indicated by broken lines. (B) Dextral view of a brain which has been dissected free. (C) Schematic frontal section plan of brain. The flexures of the brain are supposed to have been straightened before the section was cut.

adjacent brain regions are recognized. The division between telencephalon and diencephalon is an imaginary line drawn from the velum transversum to the recessus opticus . *Velum transversum* is the name given to the internal ridge formed by the deepening of the dorsal constriction which was first noted in chicks of 55 hours as indicating the impending division of the primary forebrain. The *recessus opticus* is a transverse furrow in the floor of the brain which in the embryo leads on either side into the lumina of the optic stalks. Because it is located just rostral to the optic

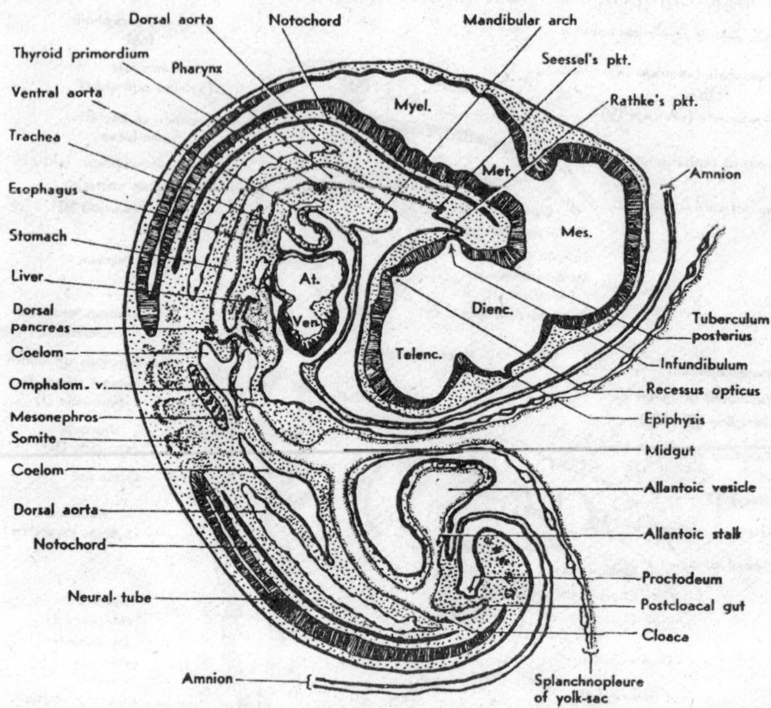

Figure 7.4 : Diagram of median longitudinal section of 4-day chick. Because of a slight bend in the embryo the section is parasagittal in the middorsal region, but for the most part it passes through the embryo in the sagittal plane.

chiasma where some of the optic nerve fibers cross, it is often spoken of as the *preoptic recess*.

The Diencephalon

The lateral walls of the diencephalon at this stage show little differentiation except ventrally where the optic stalks merge into the walls of the brain. The development of the epiphysis as a median evagination in the roof of the diencephalon has already been mentioned. Except for some enlargement, it does not differ from its condition when first formed in embryos of about 55 hours. The infundibular depression in the floor of the diencephalon has become appreciably deepened and lies in close proximity to Rathke's pocket with which it is destined to fuse in the formation of the hypophysis. Later in development the lateral walls of the diencephalon become greatly thickened to form the thalami, thus reducing the size and changing the shape of the diocoele, which is known in adult anatomy as the third brain ventricle. The anterior part of the roof of the diencephalon remains thin and becomes richly vascular. Later these vessels, invaginating the roof with them, push into the third ventricle to form the anterior choroid plexus.

The boundary between the diencephalon and the mesencephalon is an imaginary line drawn from the internal ridge formed by the original dorsal constriction between the primary forebrain and midbrain, to the tuberculum posterius. The *tuberculum posterius is* a rounded elevation in the floor of the brain, of importance chiefly because it is regarded as marking the boundary between diencephalon and mesencephalon.

The Mesencephalon

The mesencephalon as yet shows no specializations, beyond a thickening of its walls. The dorsal walls of the mesencephalon later increase rapidly in thickness and become the *corpora quadrigemina* of the adult brain. As the name implies, these are four symmetrically placed elevations. The anterior pair *(superior colliculi)* constitute the brain-center for visual reflexes; the posterior pair *(inferior colliculi)* are the center for auditory reflexes. The floor of the mesencephalon also becomes greatly thickened and is known in the adult as the *crura cerebri.* It serves as the main pathway of the fiber tracts which connect the cerebral hemispheres with the posterior part of the brain and the spinal cord. The originally capacious mesocoele is thus reduced by the thickening of the walls about it to a narrow canal, *the cerebral aqueduct* or *aqueduct of Sylvius.*

The Metencephalon

The boundary between the mesencephalon and metencephalon is indicated by the original inter-neuromeric constriction which separated them at the time of their establishement. The caudal boundary of the metencephalon is not definitely defined. It is regarded as being located approximately at the point where the brain roof changes from the thickened condition characteristic of the metencephalon to the thin condition characteristic of the myelencephalon. The metencephalon shows practically no differentiation in 4-day chicks. Later there is ventrally and laterally an extensive ingrowth of fiber tracts giving rise to the pons and to the cerebellar peduncles. The roof of the metencephalon undergoes extensive enlargement and becomes the cerebellum of the adult brain, the coordinating center for complex muscular movements.

The Myelencephalon

The dorsal myelencephalic wall is reduced in thickness, indicative of its final fate as the thin roof of the medulla. It later receives a rich supply of small blood vessels which, carrying the roof with them, grow into the myelocoele to form the posterior choroid plexus. The ventral and lateral myelencephalic walls become the floor and side walls of the medulla. Functionally the medulla serves both as a conduction path between cord and brain and as a reflex center for involuntary activities such as breathing.

The Ganglia of the Cranial Nerves

In the brain region, cells derived from the cephalic portion of the neural crest have become aggregated to form ganglia. The largest and the most clearly defined of the ganglia present in 4-day chicks is the *semilunar (Gasserian) ganglion* of the fifth (trigeminal) cranial nerve. It lies ventro-laterally, opposite the most anterior neuromere of the myelencephalon. From its cells sensory nerve fibers grow mesiad into the brain and distad to the facial and oral regions. In four-day chicks the beginning of the *ophthalmic division* of the fifth nerve extends from the ganglion toward the eye, and the beginning of the *mandibulo-maxillary division* is growing toward the angle of the mouth. Immediately cephalic to the auditory vesicle is a mass of neural crest cells which is the primordium of the ganglia of the seventh and eighth nerves. The separation of this double primordium to form the geniculate ganglion of the seventh nerve and the acoustic ganglion of the eighth nerve begins during the fourth day. Posterior to the auditory vesicle the superior ganglion of the ninth nerve can be clearly seen even in whole-mounts. The ganglia of the tenth (vagus) nerve can be recognized in sections of chicks at the end of the fourth day but are difficult to make out in whole-mounts. They show most clearly when reconstructed by the wax-plate method.

The Spinal Cord

The spinal cord region of the neural tube when first established exhibits a lumen which is elliptical in cross section. As development progresses, the lateral walls of the cord become greatly thickened in contrast with the dorsal and ventral walls which remain thin. In this process the lumen (central canal) becomes compressed laterally until it appears in cross section as little more than a vertical slit. The thin dorsal wall of the tube is known as the *roof plate;* the thin ventral wall, as the *floor plate;* and the thickened side walls, as the *lateral plates.*

The Spinal Nerves

During the fourth day the establishing of the spinal nerve roots has begun. The growth of nerve fibers from the neuroblasts can be traced only with the aid of special methods of staining. The more general steps in the development of the roots of the spinal nerves can, however, be followed in sections prepared by the ordinary methods.

In the adult each spinal nerve is connected with the cord by two roots, *a dorsal root,* which is a pathway for sensory (afferent) nerve fibers, and a *ventral root,* which is a pathway for motor (efferent) nerve fibers. Lateral to the cord the dorsal and ventral roots unite. The *spinal ganglion (dorsal root ganglion) is* located on the dorsal root between the spinal cord and the point where dorsal and ventral roots unite. Distal to the union of dorsal and ventral roots is a branch, the *ramus communicans,* which extends ventrad to a ganglion of thc sympathetic nerve cord.

When first formed from the neural crest cells, the spinal ganglion has growth of nerve fibers from cells of the spinal ganglion mesiad into the dorsal part of the lateral plate of the cord. At the same time fibers grow distad from these cells to form the peripheral part of the nerve. The fibers which arise from the dorsal root ganglion conduct sensory impulses toward the cord.

Coincident with the establishment of the dorsal root, the ventral root is formed by fibers which grow out from cells located in the ventral part of the lateral plate of the cord. Most of the fibers, which thus arise from cells in the cord and pass out through the ventral root, conduct motor impulses from the brain and cord to the muscles with which they are associated peripherally.

The sympathetic ganglia arise from cells of the neural crest, and, according to some investigators, also in part from cells moving out from the spinal cord, which migrate ventrally and form masses lying on either side of the midline at the level of the dorsal aorta. By the end of the fourth day the primordia of the sympathetic ganglia have become interconnected longitudinally by slender fibrocellular strands. The developing sympathetic ganglia then appear as local enlargements on paired cord-like structures, the

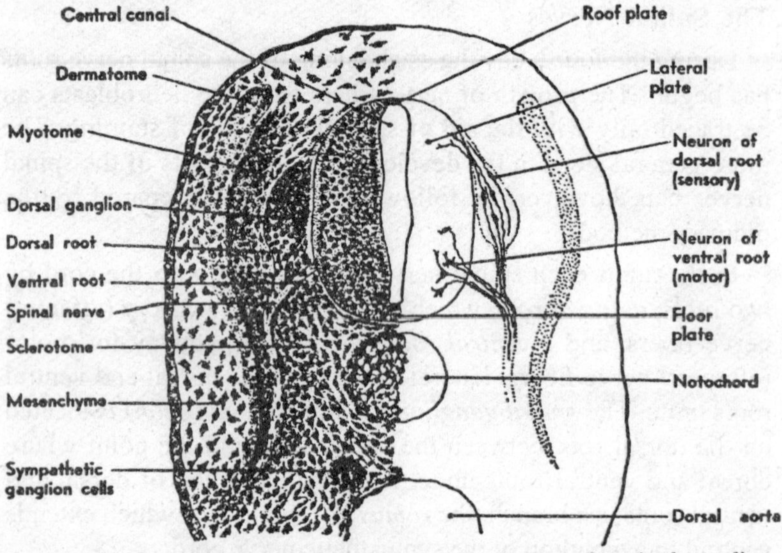

Figure 7.5 : Drawing to show the structure and relations of a spinal ganglion and the roots of a spinal nerve. The left half of the drawing represents structures as they appear after treatment by the usual nuclear staining method. The right half of the section shows, schematically, the nerve cells and the fibers growing out from them as they may be demonstrated by the Golgi method.

prevertebral sympathetic chains. Each sympathetic ganglion is connected with the corresponding spinal nerve by a fibro-cellular cord which is the primordium of the ramus communicans. Later, both sensory and motor fibers appear in the rami communicantes, putting the sympathetic ganglia in communication with the spinal nerve roots. From certain of the sympathetic ganglia, cells migrate still farther ventrad, establishing the primordia of the splanchnic sympathetic system.

The manner in which the processes of these young nerve cells grow is interesting to watch in tissue cultures. From the peripheral cytoplasm of the neuroblast a slender sprout called a *cone of growth is* formed. If this outgrowth is followed, it can be seen to advance by ameboid movements, tending to migrate along anything that furnishes a favorable substratum on which it may move. Thus the cone of growth rapidly pulls out behind itself a long slender cytoplasmic filament which is the young nerve fiber. If it encounters any resistant substance such as a bit of young cartilage in the explanted tissue or a particularly dense portion of the plasma clot, its growing tip will turn aside into paths offering less resistance.

These tendencies of a growing nerve fiber to follow along something which offers it a substratum and to be turned aside by obstacles are vitally important in dealing with the repair of nerve injuries. When a nerve is cut, the parts of the fibers which are severed from the nerve cells die. When regeneration starts, it begins with the cut ends of the parts of the fibers which remained in connection with the nerve cells. The cut ends send out cones of growth similar to those arising from embryonic neuroblasts. If the injury is a clean cut, so the proximal end of the nerve can be accurately sewed into place against the degenerated distal end, the regenerating fibers will follow along the old nerve sheaths as a pathway and eventually reestablish their peripheral connections. If, however, the approximation is not a clean one, scar tissue turns aside the regenerating nerve tips, so they do not find their way into their old pathway. If an injury is so extensive that it leaves a gap between the living ends of the nerve fibers and their former sheaths, the regenerating fibers will become hopelessly entangled in the new intervening connective tissue unless they are provided with an artificial pathway. A bit of frozen and vacuum-dried nerve may be used for this purpose. Carefully fixed across the gap in the interrupted nerve, its parallel sheaths form a path by which the regenerating nerve fibers may grow across and pick up their own old pathways. In such nerve grafts, of course, the grafted tissue serves merely as a temporary bridge and is rapidly resorbed and replaced by new host tissue.

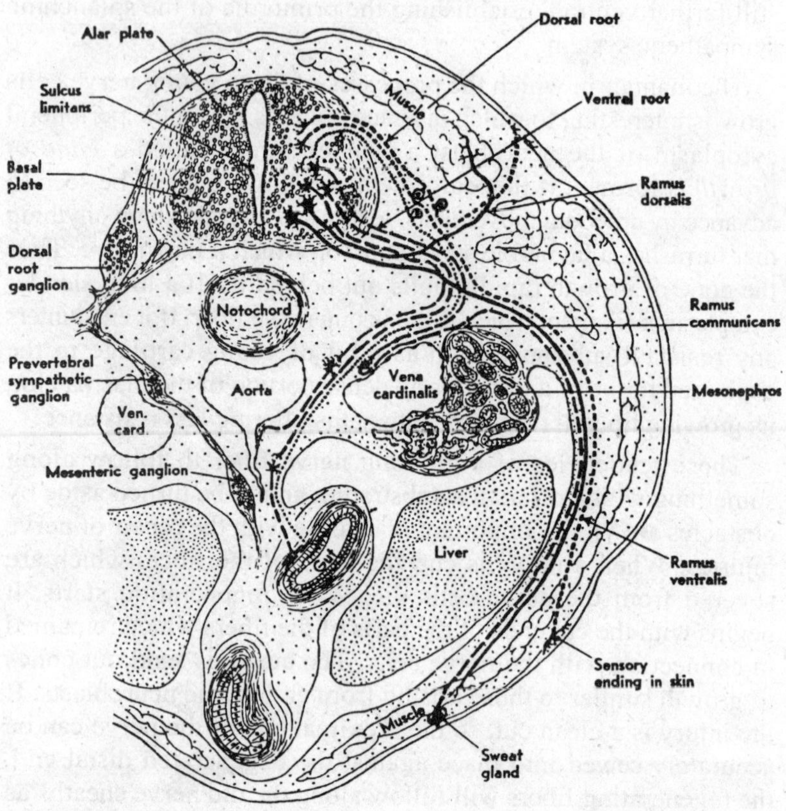

Figure 7.6 : Schematic diagram indicating the various connections made by the neurons which develop in a typical spinal nerve.

The neurologist classifies the fibers in a spinal nerve according to their relations and functions. The components of a typical spinal nerve on this basis are:

I . ***Afferent :***

A. General Somatic Afferent : **1. Fxteroceptive, i.e.,** ***fibers conducting impulses from the external surface of the body such as touch, pain, temperature. (Represented in this figure by short broken lines.)*** **2. Proprioceptive, i.e.,** ***fibers carrying impulses of position sense from joints, tendons, and muscles. (Not represented in this diagram.)***

B. General Visceral Afferent : Fibers from viscera (interoceptive) by way of sympathetic chain

The lens, meanwhile, arises as a thickening of the superficial ectoderm which becomes depressed to form a vesicular invagination extending into the optic cup.

In chicks of four days the choroid fissure has become narrowed by the growth of the walls of the optic cup on either side of it. At the same time the orifice of the optic cup becomes narrowed by convergence of its margins toward the lens. Meanwhile the lens has become freed from the superficial ectoderm and forms a completely closed vesicle. Sections of the lens at this stage show that the cells constituting that part of its wall which lies toward the center of the optic cup are becoming elongated to form the *lens fibers*.

The Coulombres (1993) have carried out experiments on the developing lens in chick embryos, which give important information as to the factors controlling the elongation of lens fibers. With superb technical skill they removed the lens from the optic vesicle on the fifth day and replanted it with the lens epithelium facing toward the retina instead of its normal position toward the cornea. In such reversed lenses the lens fibers which had started to elongate ceased their growth, and the lens epithelium, which normally does no such thing, began to form a new set of lens fibers. This clearly indicates that the formation of lens fibers is not simply a matter of the age of the formative cells but that, taken at this early stage, the lens can respond to a changed environment by complete reversal of its polarity.

At this stage we can identify the beginning of most of the structures of the adult eye. The thickened internal layer of the optic cup will give rise to the *sensory layer o f the retina* (Fig. 7.6, *B)*. Fibers arise from nerve cells in this layer of the retina and grow along the groove in the ventral surface of the optic stalk toward the brain to form the optic nerve. The external layer of the optic cup gives rise to the *pigment layer of the retina*. Mesenchymal cells can be seen aggregating in progressively increasing numbers about the outside of the optic cup. From these the *sclera* and *choroid coat* are derived. Some of the mesenchyme makes its way into the optic cup through the choroid fissure and

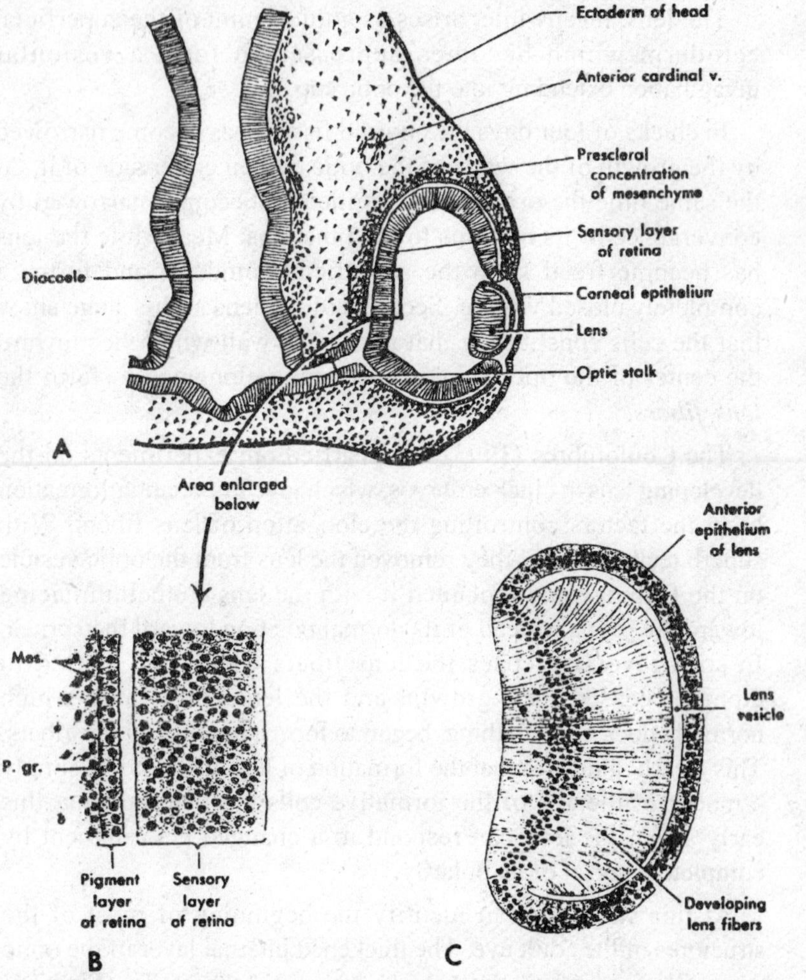

Ectoderm of head
Anterior cardinal v.
Prescleral concentration of mesenchyme
Sensory layer of retina
Diocoele
Corneal epithelium
Lens
Optic stalk
A
Area enlarged below
Mes.
P. gr.
Pigment layer of retina
Sensory layer of retina
B
Anterior epithelium of lens
Lens vesicle
Developing lens fibers
C

Figure 7.7 : Drawings to show structure of the eye of a 4-day chick. (A) Diagram graphy of eye region. (B) Drawing to show cellular or organization of the pigment and show (C) Drawing to show cellular organization of the lens. Abbreviations: Mes., mesenchymal cell; P. gr., pigment granule.

to the cellular elements of the *vitreous body*. The complex *ciliary apparatus* of the adult eye is derived from the margins of the optic cup adjacent to the lens. The *corneal and conjunctival epithelium* arise from the superficial ectoderm overlying the eye. Mesenchymal cells which make their way between the lens and the corneal epithelium give rise to the *substantia propria* of the cornea.

The Ear

Of the structures taking part in the formation of the ear, the first to appear is the auditory placode. The auditory placode is recognizable in 36-hour chicks as a thickened plate of ectoderm. Almost as soon as it appears, the placode sinks below the level of the surrounding ectoderm to form the floor of the auditory pit. By constriction of its opening to the surface, the epithelium of the auditory pit becomes separated from the ectoderm of the head and comes to lie close to the lateral wall of the myelencephalon. A tubular stalk, the *endolymphatic duct,* remains for a time adherent to the superficial ectoderm, marking the location of the original invagination.

The degree of development reached by the ear primordium in 4-day chicks gives little indication of the nature of the later processes by which the ear is formed. The auditory vesicle by a very complex series of changes will give rise to the entire epithelial portion of the *internal ear* mechanism. Nerve fibers arising from the *acoustic ganglion* grow into the brain proximally, and to the internal ear distally, establishing nerve connections between them. There is at this stage no indication of the differentiation of the external auditory meatus. The dorsal and inner portion of the hyomandibular cleft which gives rise to the *auditory (Eustachian) tube* and to the middle ear chamber has not yet become associated with the auditory vesicle.

The Olfactory Organs

The olfactory organs are represented in 3- and 6-day chicks by a pair of depressions in the ectoderm of the head. These so-called *olfactory pits* are located ventral to the telencephalic vesicles and just interior to the mouth. By growth of the processes which surround them, the olfactory pits become greatly deepened. The epithelium ining the pits eventually comes to lie at the extreme upper part of the *nasal chambers* and constitutes the *olfactory epithelium.* Nerve fibers grow from these cells to the telencephalic lobes of the brain to form the *olfactory nerves.*

Summary of Development Prior to the Third Day

The primary entoterm which gives rise to the epithelial lining of the digestive and respiratory systems and their associated glands becomes established as a separate layer before the egg is laid. In its early relationships the entoderm is a sheet-like layer of cells lying between the ectoderm and the yolk and attached peripherally to the yolk, so the primitive gut cavity is bounded only dorsally by the entoderm and ventrally has the yolk as a temporary floor

Only the part of the entoderm which lies within the embryonal area is involved in the formation of the enteric tract. The peripheral portion of the entoderm goes into the formation of the yolk-sac. There is at first no definite line of demarcation between the entoderm destined to be incorporated into the body of the embryo and that which remains extra-embryonic in its associations. The foldings which appear later, separating the body of the embryo from the yolk, establish for the first time the boundaries between intra-embryonic and extra-embryonic entoderm.

The first part of the gut to acquire a complete entodermic lining is the foregut. Its floor is formed by the caudally progressing concrescence of the entoderm which takes place as the subcephalic and lateral body folds undercut the cephalic part of the embryo. At a considerably later stage the hindgut is formed in a similar manner by the progress of the subcaudal fold toward the head. Between the foregut and the hindgut, the midget remains open to the yolk ventrally. As the embryo is more completely separated from the yolk, the foreget and hindget increase in extent at the expense of the midget. By the fourth day of incubation the midget is reduced to the region where the yolk-stalk opens into the enteric tract.

Establishing the Oral Opening

As we saw in the study of younger chicks, when the gut is first established it ends as a blind pocket both cephalically and caudally. In embryos of 55 to 60 hours the processes leading toward the establishment of the oral opening were, however, clearly indicated. A midventral evagination of the pharynx had been established immediately cephalic to the mandibular arch. Opposite this outpocketing of the pharynx and growing in to meet it, the stomodaeal depression had been formed. The only bar to an actual opening was the oral plate, a thin membrane constituted by the apposition of the pharyngeal entoderm and the stomodaeal ectoderm. The communication of the foregut with the outside is finally established, during the third day, by the breaking through of the oral plate. Following the rupture of the oral plate, growth of the surrounding structures rapidly deepens the originally shallow stomodaeal depression. The region where the oral plate was originally located in the embryo eventually becomes, in the adult, the region of transition from oral cavity to pharynx.

The formation of the oral opening in the manner described does not take place at the extreme anterior end of the foregut. A small gut pocket extends cephalic to the mouth. This so-called pre-oral gut rapidly becomes less conspicuous after the breaking through of the oral plate. The small depression which in older embryos marks its location is known as Seessel's pocket. Even this small depression eventually disappears altogether. Its importance lies wholly in the fact that its margin indicates for some. time the place at which ectoderm and. entoderm originally became continuous in the formation of the oral opening.

The Pharynx

Caudal to the oral opening the foregut has become flattened dorso-ventrally and widened laterally to form the pharynx. This is a region we have already become acquainted with in dealing with younger embryos, but its relations are so important that some review of them here will not be amiss. On either side the pharyngeal lumen shows a series of extensions or bays known as the *pharyngeal pouches*. Each pharyngeal pouch lies opposite an external gill furrow or *branchial groove*. This leaves in these areas only a thin layer of tissue separating the pharyngeal lumen from the outside. This layer is composed internally of pharyngeal entoderm and externally of the ectoderm of the bottom of the branchial groove. There may or may not be a small amount of mesenchyme between these two epithelial layers of the *gill plate*. Sometimes the more cephalic gill plates actually break through, establishing transitory open gill clefts.

Between adjacent gill grooves or clefts the lateral walls are greatly thickened and filled with closely packed mesenchymal cells. These thickened areas are known as the *branchial arches* (also sometimes called visceral arches or gill arches). One of the most important relationships to grasp in the study of young embryos is the manner in which the *aortic arches* lie embedded in the tissue of the branchial arch of corresponding number. The stereogram was planned especially to emphasize these relationships. It should be intensively studied in conjunction with the section diagrammed below it until a clear three-dimensional concept of the relations in this region has been grasped.

The Pharyngeal Derivatives

Several structures, which do not become parts of the digestive system, arise in the pharyngeal region. Nevertheless the origin of their epithelial portions from foregut entoderm and their early association with this part of the gut tract make it convenient to consider them in connection with the digestive system.

The *thyroid gland* arises from the floor of the pharynx as a median diverticulum which makes its appearance at a cephalo-caudal level between the first and second pair of pharyngeal pouches. Toward the end of the fourth day the thyroid evagination has become sacral. Later that in that group also, they probably do not give rise to thyroid tissue as formerly believed, in spite of the very close positional relation which comes to exist between them and the median thyroid evagination.

The two pairs of *parathyroid glands,* also, arise as bud-like outgrowths from the pharynx. One pair of parathyroids is budded off the caudal faces of the third pharyngeal pouches, and the other pair arises in a similar manner from the fourth pouches. These parathyroid primordia are, however, not usually recognizable until later stages of development than those here under consideration.

The *thymus* of the chick is barely indicated, if present at all, on the fourth day of incubation. It takes its origin primarily from diverticula arising from the posterior faces of the third and fourth pharyngeal pouches. The original epithelial character of the thymus is soon largely lost in an extensive ingrowth of mesenchyme, and the organ becomes chiefly lymphoid in its histologic characteristics.

The Trachea

The first indication of the formation of the respiratory system appears in 3-day chicks as a midventral groove in the pharynx. Beginning just posterior to the level of the fourth pharyngeal pouches and extending caudad, this *laryngo-tracheal groove* deepens rapidly and by closure of its dorsal margins becomes separated from the pharynx except at its cephalic (laryngeal) end. The tube thus formed is the *trachca*, and the opening which pcrsists between the laryngeal end of the trachea and the pharynx is the *glottis*. The original entodermal evagination gives rise only to the epithelial lining of the trachea, the supporting structures of the tracheal walls being derived from the surrounding mesenchyme.

The Lung-buds

The tracheal evagination grows caudad and bifurcates form a pair of lung-buds. As the lung-buds develop they grow into the ose mesenchyme on either side of the midline. This relationship of mesenIyme to growing entodermal epithelium, as seen here in the lungs, and so in the developing gastrointestinal tract, is an interesting one. Apparently the formation of tubular structures, such as the bronchi of the lungs or the acts and secreting tubules of digestive glands, is dependent on the presence mesoderm about the growing entoderm. If entoderm is grown in a culture without the presence of any mesoderm, it tends to spread out as a flat, zed sheet of sprawling cells. If, however, mesenchymal cells are added to such cultures, the entodermal epithelium forms tubular structures. These bular structures, moreover, will take on the characteristics of the lung. or a digestive gland, or the epithelial lining of the gut tract according to the primordial region from which the original entodermal cells were taken. Thus again, this time, at the level of tissue differentiation, we see the influence of one growing part of the embryo on another.

In the caudo-lateral growth of the primordial lung-buds the surrounding splanchnic mesoderm is pushed ahead of them and comes to constitute their outer investment. The entodermal buds give rise only to the epithelial lining of the bronchi and the air passages and air chambers of the lungs. The connective tissue stroma of the lungs is derived from mesenchyme immediately surrounding the entodermal buds, and their pleural covering is formed from their primary investment of splanchnic mesoderm.

The Esophagus and Stomach Immediately caudal to the glottis is a narrowed region of the foregut, which becomes the esophagus, and farther caudally a slightly dilated region, which becomes the stomach. The concentration of mesenchymal cells about the entoderm of the esophageal and gastric regions foreshadows the formation of their muscular and connective tissue coats.

The Liver

In all vertebrates the liver arises as a diverticulum from the ventral wall of the gut, a little caudal to the stomach. In chick embryos the hepatic primordium can first be recognized at about the 22-somite stage. It appears just as the part of the gut from which it arises is acquiring a floor by the concrescence of the margins of the anterior intestinal portal. As a result the evagination which is the primordium of the liver is located for a short time on the lip of the intestinal portal and grows cephalad toward the confluence of the omphalomesenteric veins just before they enter the sinus venosus. As closure of the gut floor is completed the hepatic diverticulum comes to lie in its characteristic position in the ventral wall of the gut. In embryos of 4 days the original evagination has grown out in the form of branching cords of cells and become quite extensive in mass. In its growth the liver pushes ahead of itself the splanchnic mesoderm which surrounds the gut, with the result that the liver from its first appearance is invested by mesoderm.

The proximal portion of the original evagi.nation remains open to the intestine, and serves as the duct of the liver. This primitive duct later undergoes regional differentiation and gives rise in the adult to the *common bile duct,* to the *hepatic and cystic ducts,* and to the *gallbladder.* The cellular cords which bud off from the diverticulum become the secretory units of the liver (*hepatic tubules).*

The same process of concrescence which closes the floor of the foregut involves the proximal portion of the omphalomesenteric veins which, when they first appear, lie in the lateral folds of the anterior intestinal portal. As the intestinal portal moves caudad in the lengthening of the foregut, the proximal portions of the omphalomesenteric veins are brought together in the midline and become fused. The fusion extends caudad nearly to the level of the yolk-stalk. Distal to this point they retain their original paired condition. In its growth the liver surrounds the fused portion of the omphalomesenteric veins. This early association of the omphalomesenteric veins with the liver foreshadows the way in which blood channels coming in from the vitelline vascular plexus are to be involved in the establishment of the hepatic-portal circulation of the adult.

The Pancreas

The pancreas is derived from evaginations appearing in the walls of the intestine at the same level as the liver diverticulum. There are three pancreatic buds, a single median dorsal, and a pair of ventro-lateral buds. The dorsal evagination appears at about 72 hours, the ventro-lateral evaginations toward the end of the fourth day. The dorsal pancreatic bud arises from gut entoderm directly opposite the liver diverticulum and grows into the dorsal mesentery. The ventrolateral buds arise close to the point where the duct of the liver connects with the intestine so that the ducts of the liver and the ventral pancreatic ducts open into the intestine by a common duct *(ductus choledochus)*. Later in development the masses of cellular cords derived from the three pancreatic primordia grow together and become fused into a single glandular mass, but, in birds, usually two and in rare cases all three of the original ducts persist in the adult.

The Midgut Region

In chicks of four days the enteric tract shows no local differentiation from the level of the liver to the cloaca except where the yolk-sac is attached. All of the gut tract between the stomach and the yolk-stalk, and the anterior fourth of the gut which lies caudal to the yolkstalk, are destined to become the small intestine. The posterior two-thirds of the hindgut becomes large intestine and cloaca.

The Cloaca

The beginning of the formation of the cloaca is indicated in chicks of four days incubation by a dilation of the posterior portion of the hindgut. Although extensive differentiation in the cloacal region does not appear until later in development, certain of its fundamental relationships are established at this stage.

The cloaca of an adult bird is the common chamber into which the intestinal contents, the urine, and the products of the reproductive organs are received for discharge. The first appearance of the cloaca in the embryo as a dilated terminal portion of the gut establishes at the outset the relations of cloaca and intestine that persist in the adult.

Although the urinary system is not at this stage developed to conditions which resemble those in the adult, the parts of it which have been established are already definitely associated with the cloaca. The proximal portion of the allantoic stalk, which is the homolog of the urinary bladder of mammals, opens directly into the cloaca. When the urinary system of the embryo is considered, we shall see that the ducts which drain the developing excretory organs also open into the cloacal region on either side of the allantoic stalk.

There is at this stage but little indication of the formation of the gonads. The relation of the sexual ducts to the cloaca can be made out only by the study of older embryos.

The Proctodaeum and the Cloacal Membrane

Indications of the formation of the cloacal opening to the outside appear during the fourth day of incubation. Its establishment is accomplished in much the same manner as the establishment of the oral opening. A ventral outpocketing of the hindgut arises just caudal to the point at which the allantoic stalk opens into the cloaca. At the same time a depression appears in the overlying ectoderm. The external depression which grows in toward the gut pocket is known as the *proctodaeum.* The double epithelial layer formed by the meeting of gut entoderm with proctodaeal ectoderm is the cloacal membrane. The formation of the proctodaeum and the cloacal membrane clearly indicates the location of the future cloacal opening although an open communication is not established by the rupture of the cloacal membrane until considerably later. The cloacal opening does not form at the extreme posterior end of the hindgut and there is, therefore, a post-anal pocket of the hindgut suggestive of the pre-oral pocket of the foregut.

THE CIRCULATORY SYSTEM

Interpretation of the Embryonic

Circulation The embryonic circulation is difficult to understand only when the meaning of its arrangement is overlooked. If one bears in mind the significance of the circulatory system in organic economics and the fact that any embryo must go through certain ancestral phases of organization before it can arrive at its adult structure, the changes in the arrangement of vascular channels during the course of development form a coherent and logical story.

In the embryo, as in the adult, the main vascular channels lead to and from the centers of metabolic activity. The circulating blood carries food from the organs concerned with its absorption to parts of the body remote from the source of supplies. It also carries oxygen to all the tissues of the body, from organs which are especially adapted to facilitate the taking of oxygen into the blood; and waste materials, from the places of their liberation, to the organs through which they are eliminated. One of the primary reasons the arrangement of the vessesl in an embryonic bird or mammal differs so much from that in the adult is the fact that the embryo lives under conditions totally unlike those which surround its parents. Its centers of metabolic activity are, therefore, different; and, since the course of its main blood vessels is determined by these centers, the vascular plan is different. No such profound changes as we find in birds and mammals occur between the embryonic and the adult stages in the circulation of a fish where embryo and adult are both living under similar environmental conditions.

The organs which in the adult carry out such functions as digestion; and absorption, respiration, and excretion are extremely complex and highly differentiated. They are for this reason slow to attain their definitive condition and are not ready to become functional until toward the close of the embryonic period. Moreover the conditions which surround certain of the developing organs during embryonic life would prevent their becoming functional even were they sufficiently developed so to do. Suppose the lungs, for example, were functionally competent at an early stage of development. The fact that the embryo is reliving ancestral conditions submerged in the amniotic fluid renders its lungs as incapable of functioning as those of a man under water. Likewise the developing stomach of a chick encased in its shell can receive no raw food materials by way of the mouth. Further examples are not necessary to make it obvious that, were the embryo dependent on the same organs which carry on metabolism in the adult, its development would be at an impasse.

Nevertheless an embryo must solve the problem of existence during the time in which it is building up a set of organs similar to those of its parents. The chick must have not only the raw food material supplied by its mother in the form of yolk, it must have in addition a means of digesting the yolk, absorbing it, and carrying it to the places where it can be utilized. Furthermore the utilization of food material to produce the energy expressed in growth depends on the presence of oxygen. So for growth there must be a means of securing oxygen and carrying it, as well as food, to all parts of the body. Nor can continued growth go on unless the waste products liberated by the growing tissues are eliminated. At the outset of its development the embryo must, therefore, establish organs for the digestion and absorption of food, the securing of oxygen, and the elimination of waste products. In various types of embryos these functions are carried out by the yolk-sac or the allantois or by both together, in a manner depending on the exigencies of the embryo's living conditions. These structures are relatively simple in comparison with the organs which carry out the corresponding fuctions in the adult, but their activities are so essential and so extensive that the vascular channels supplying them are a dominating feature in the embryonic circulation.

While main circulatory channels are thus always established in relation to centers of metabolic activity, we find in many situations in the embryo an unmistakable phylogenetic impress on the manner of their establishment or on the details of their arrangement. Take, for example, the aortic arches. The blood leaving the ventrally located heart must pass around the gut to reach the dorsally located aorta. In fishes six pairs of aortic arches encircle the pharynx, breaking up in the gills into capillaries which carry out the indispensable function of oxygenating the blood. Once an animal has replaced its gills with lungs, it makes little difference functionally whether or not the blood passes by each gill cleft on its way from heart to aorta. In adult birds and mammals we find this communication simplified to a single main aortic arch. But in the embryos of both birds and mammals the whole series of symmetrical aortic arches appear for a time, encircling the pharynx and passing in close relation to vestigial gill clefts. This can be interpreted only as a recapitulation of ancestral conditions-conditions which, although they have ceased to be of functional importance, appear nevertheless as a developmental phase on the way to a more highly differentiated plan of adult structure. Our interpretation of the aortic arches, then, would take into consideration first the fundamental functional necessity of a channel connecting the ventrally located heart with the dorsally located aorta. Second, looking at the striking arrangement of the series of aortic arches in relation to the gill arches and clefts, one sees a repetition of the structural plan which existed in water-living ancestral forms. Thus in the end we are led back again to functional significance, for the relationship of the aortic arches to the gill arches writes into the story of individual development an unequivocal record of the evolutionary phase when the gills were a center of primary metabolic importance.

Applied to the development of the circulatory system as a whole, this tendency to repeat phyletic history means that the earliest form in which the circulatory mechanism of a bird or a mammal appears is patterned after a more primitive type and, therefore, cannot be a miniature of adult conditions. The simple tubular heart pumping blood out over aortic arches to be distributed over the body by the aorta and returned to the posterior part of the heart by a bilaterally symmetrical venous system, in short the foundational vascular plan which we see in sauropsidan embryos, is essentially the plan of the circulation in fishes. When we realize this, we are puzzled neither by the early appearance of a full complement of aortic arches nor by their subsequent disappearance to make way for a new respiratory circulation in the lungs. Starting from a logical beginning in simpler ancestral conditions, we see the march of progress toward the consummation of embryonic life with the attainment of an organization like that of the immediate parent. And at each turning point the direction of further advance is sharply limited. Development can proceed only along lines that permit the maintenance of an efficient metabolism, both in the temporary organs characteristic of embryonic life and in the slowly developing organs which must be prepared to meet adult living conditions.

Finally, we must bear in mind that only relatively late in development, as the various organs become established in their adult relationships, can we expect to see the emergence of the vascular plan characteristic of the adult. Following the primitive embryonic stage there must be transitional phases when functioning embryonic channels are present side by side with channels being prepared to take over their activity in the adult. For all changes in the circulatory system of a living organism must be gradual. Any changes which were sufficiently abrupt to interfere with the circulation would result in disaster for the embryo. Even slight curtailment of the normal blood supply to any region would cause its growth to cease; any marked local decrease in the circulation would result in local atrophy or malformation; complete interruption of any important circulatory channel, even for a short time, would inevitably mean the death of the embryo.

Main Routes of the Embryonic Circulation

Before taking up the various parts of the circulatory system in detail it might be well to consider briefly the main routes of the embryonic circulation. The circulation of young chick embryos involves three main arcs of which the heart is the common center and pumping station. One of these circulatory arcs, the vitelline, carries blood to the yolk-sac where food materials are absorbed and then returns the food-laden blood to the heart for distribution within the embryo. Another arc carries blood to and from the allantois. The distal portion of the allantois lies close beneath the egg shell, and the blood circulating in the allantoic vessels is thereby brought into a location where interchange of gases can be carried on with the air which penetrates the shell. It is in the allantoic circulation that the blood gives off its carbon dioxide and acquires a fresh supply of oxygen. Thc allantoic circulation is also the embryo's means of eliminating nitrogenous waste material from the blood. The remaining circulatory arc is confined to the body of the embryo. The intraembryonic circulation has many distributing and collecting vessels, but all of them are alike in function in that they bring food material and oxygen to, and carry waste material from, the various parts of the developing body. Nowhere in their course are the vessels of the intra-embryonic circulation involved in adding food material or oxygen to that already contained in the blood they convey, and nowhere do they free the blood from waste materials until well along in development, when the nephroi become functional.

In the heart the blood from the three circulatory arcs is mingled. As it leaves. the heart the mixed blood is not as rich in food material as the blood coming in through the omphalomesenteric veins, nor as free from waste materials and as rich in oxygen as the blood returned over the allantoic veins.' Its condition of serviceability to the embryo is, however, constantly maintained at a good average by the incoming vitelline and allantoic blood.

The Vitelline Circulation

The earliest indication of blood and blood vessel formation is at the chick's source of food supply. Blood islands appear in the extra-embryonic splanchnopleure of the yolk-sac toward the end of the first day of incubation and rapidly become differentiated to form vascular endothelium enclosing central clusters of primitive blood corpuscles. By extension and anastomosing of neighboring islands, a plexus of blood channels is formed in the yolk-sac. Further extension of the vitelline plexus brings it into communication with the omphalomesenteric veins which have been developed in the embryo as caudal extensions of the endocardial primordia.

Toward the end of the second day of development the omphalomesenteric arteries establish communication between the dorsal aortae and the vitelline plexus. There is now a system of open channels leading from the embryo to the yolk-sac, and back again to the embryo. With the completion of these channels the heart, which has for some hours been building up the efficiency of its pulsation, is able to set the blood in circulation. It is thus, as we have seen, at about the 40th hour of incubation that the blood cells formed in the yolk-sac are for the first time carried into the body of the embryo.

Circulating in the rich plexus of small vessels on the yolk, the blood finally makes its way either directly into one of the larger vitelline veins, .or into the sinus terminalis which acts as a collecting channel, and then through the sinus terminalis to one of the vitelline veins. The vitelline veins converge toward the yolk-stalk where they empty into the omphalomesenteric veins. The omphalomesenteric veins, at first paired throughout their entire length, have been brought together proximally by the closure of the ventral body-wall and become fused to form a median vessel within the body of the embryo. It is through this vessel that the blood returning from the vitelline circuit eventually reaches the heart. In the heart the blood of the vitelline, intra-embryonic, and allantoic circulations is mingled. The mixed blood passes out by the ventral aorta and the aortic arches into the dorsal aorta. Leaving the dorsal aorta through the vitelline arteries a certain portion of this blood is returned to the yolk-sac for another circuit through it.

It should not be inferred that the blood stream "picks up" deutoplasmic granules and carries them to the embryo. The acquisition of food material by the blood depends on the activities of the entodermal cells lining the yolk-sac. These cells secrete digestive enzymes which break down the deutoplasmic granules. The liquefied material is then absorbed by the yolksac cells and transferred to the blood. The blood carries the food material in soluble form to the embryo where it is finally assimilated.

The Allantoic Circulation

The allantoic arteries arise by the prolongation and enlargement of a pair of vessels arising ventrally from the aorta at the level of the allantoic stalk. Their size increases rapidly as the allantois increases in extent. From them the blood is distributed in a rich plexus of vessels which ramify in the mesoderm of the allantois.

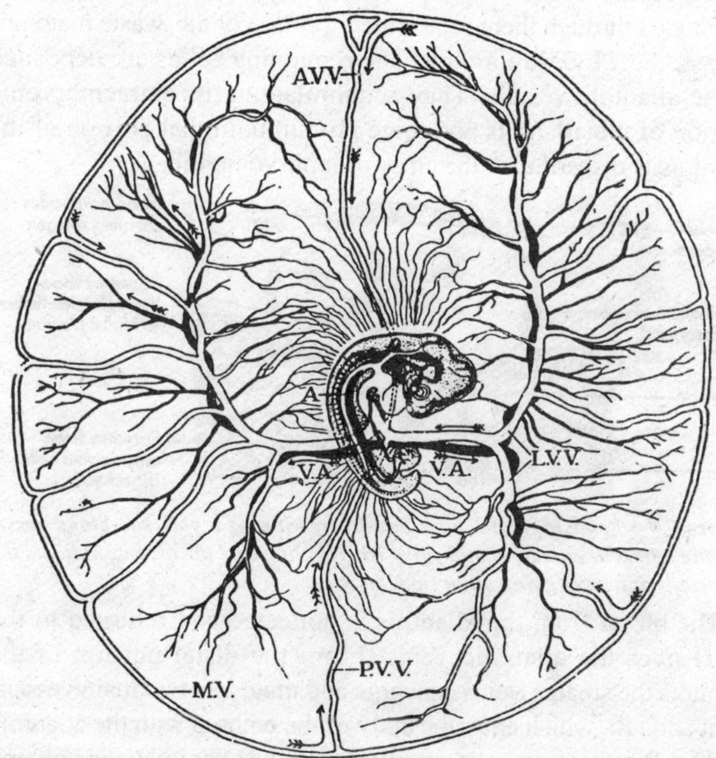

Figure 7.8 : Diagram to show course of vitelline circulation in chick of about 4 days.

Abbreviations: A, *dorsal aorta;* A. V. V., *anterior vitelline vein;* L. V. V., *lateral vitelline vein;* M. V., *marginal vein (sinus terminalis);* P. V. V., *posterior vitelline vein;* V. A., *vitelline artery. The direction of blood flow is indicated by arrows.*

The situation of the allantois directly beneath the porous shell is such that the blood can carry on interchange of gases with the outside air. It is in the rich plexus of small allantoic vessels where the surface exposure is very great that the blood gives off its carbon dioxide and takes up oxygen.

At a later stage of development the ducts of the embryonic excretory organs open into the allantoic stalk near its cloacal end. As the excretory organs become functional, the allantoic vesicle becomes the repository for the nitrogenous waste materials eliminated through them. The watery portion of the waste materials is passed off by evaporation. The remaining solids are deposited in the allantoic vesicle. They accumulate in the extraembryonic portion of the allantois and there remain until that portion of the allantois is discarded at the close of embryonic life.

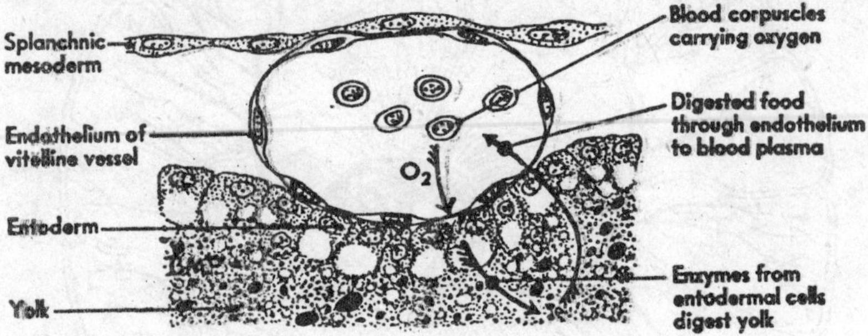

Figure 7.9 : Semischematic high-power drawing of a yolk-sac blood vessel and the adjacent entoderm and yolk. The labeling and the arrows indicate the important processes going on in such an area.

The blood from the allantois is collected and returned to the heart over the allantoic veins. From the distal portion of the allantois the smaller veins converge and unite into two main vessels, right and left, which enter the body of the embryo with the allantoic stalk. After their entrance into the body the allantoic veins extend cephalad in the lateral body-walls . They enter the sinus They enter on either side of the entrance of the omphalomesenteric vein.

The Intra-embryonic Circulation

The earliest vessels of the intra-embryonic circulation to appear .are the large vessels communicating with the heart. In chicks of 33 hours the ventral aorta leads off from the heart cephalically and bifurcates ventral to the pharynx giving rise to a 'single pair of aortic arches. These first aortic arches pass dorsad around the antero-lateral walls of the pharynx and are continued caudally along the dorsal wall of the gut as the paired dorsal aortae.

When, toward the end of the second day of incubation, branchial clefts and branchial arches appear, the original pair of aortic arches comes to lie in the mandibular arch. In each of the branchial arches posterior to the mandibular, new aortic arches will be formed connecting the ventral aortae with the dorsal aortae. By 50 hours two pairs of aortic arches are present and a third is beginning to form. In chicks of 60 hours incubation three aortic arches have been completed, and the fourth is usually starting to form. By the end of the fourth day a rudimentary fifth and a welldeveloped sixth arch are present. From their first appearance the fifth aortic arches are very small and they soon disappear altogether. The first and second pairs of aortic arches have by this time suffered a marked diminution in size, which is indicative of their final disappearance. In many embryos of this age range the first arches, and in a few the second also, have disappeared altogether. This leaves only the third, fourth, and sixth pairs of aortic arches. These arches persist intact for some time, and parts of them remain permanently, being incorporated in the formation of the aortic arch and the main vessels arising from it, and in the roots of the pulmonary arteries.

In embryos late in the second day we saw the appearance of plexiform channels extending from the first aortic arch toward the forebrain and noted that they foreshadowed the formation of the *internal carotid arteries.* By 60 hours the internal carotid artery is established as a definite trunk extending toward the brain from the point where the first aortic arch turns into the dorsal aortic root. With the regression of the first two aortic arches during the fourth day of development, the dorsal aortic root at this level is appropriated as a feeder to the original internal carotid artery, thereby making it appear to originate where the third aortic arches merge with the dorsal aortic roots. The meshwork of small vessels fed by the internal carotid artery and developing in close relation to the brain is one of extraordinary richness. In embryos of this age it is possible to see suggestions of certain of the main named branches characteristic of the adult, but for the most part they are represented only by primordial capillary plexuses.

The regression of the first two aortic arches plays a still more prominent role in the formation of the *external carotid arteries.*

When these two arches lose connection with the dorsal aortic roots, the part of the ventral aortic roots which formerly fed them persists. These vessels represent the main stems of the external carotid arteries which later develop many important branches supplying the region of the mandible and the front of the neck and the face.

In reptiles, birds, and mammals the main adult vessels which connect the heart with the dorsal aorta are derived from the fourth pair of aortic arches of the embryo. The paired condition of these arches persists as the adult condition in reptiles, but in birds and mammals one of the arches degenerates before the end of embryonic life. In birds the left arch degenerates leaving the right one as the arch of the adult aorta; in mammals the right arch degenerates leaving the left as the aortic arch of the adult aorta.

The dorsal aortae, at first paired, later become fused to form a median vessel. As we have seen, the fusion begins at cardiac level. It extends cephalad, but a short distance, never involving the region of the dorsal aortic roots where they receive the aortic arches. Caudally the aortae eventually become fused throughout their entire length.

Early in development the aorta gives rise to a segmentally arranged series of small vessels which extend into the dorsal body-wall. At the level of the anterior appendage-buds several pairs of these *dorsal segmental arteries* extend into the wing buds. Later one of these becomes enlarged to form the *subclavian arteries.*

Coincident with the development of the allantois, a pair of *ventral segmental vessels* opposite the allantoic stalk become enlarged and extend into it as the *allantoic arteries.* These, as we have seen, feed a rich plexus of small vessels which, with the growth of the allantois, come to lie close beneath the porous shell and become concerned with gaseous interchanges.

The blood supply to the posterior appendage starts as a plexus fed by several segmental vessels at the level of the appendage-bud. Later this primordial plexus comes to be fed by a branch arising from the allantoic artery. This new vessel becomes the *external iliac artery.*

In the adult, three vessels arising from the dorsal aorta supply the abdominal viscera. They are the *coeliac, superior mesenteric,* and *inferior mesenteric arteries.* In 4-day chicks these vessels are usually represented only by the omphalomesenteric arteries. The omphalomesenteric arteries arise as paired vessels, but in the closure of the ventral body-wall of the embryo they are brought together and fused to form a single vessel which runs in the mesentery from the aorta to the yolk-stalk. With the atrophy of the yolk-sac the proximal part of the omphalomesenteric artery persists as the superior mesenteric of the adult. The coeliac and the inferior mesenteric arteries arise from the aorta independently, usually at a somewhat later stage. Occasionally, however, the coeliac can be identified in 4-day embryos.

As development progresses, any arterial trunk growing into an organ will exhibit a pattern of branching which bears a definite relation to the structure of the organ it is supplying. The vascular sprouts tend to follow along the developing connective tissue framework of the organ, assuming a configuration dictated by its arrangement. This can very prettily be demonstrated experimentally by the transplantation of an organ primordium before it has become vascularized. When this is done, the host vessels which grow in and supply the transplanted organ will be found to show the same basic arrangement that would have appeared if the organ had been allowed to develop undisturbed.

The *anterior* and *posterior cardinal veins* are the principal afferent systemic vessels of the early embryo. They appear toward the end of the second day as paired vessels extending cephalically and caudally on either side of the midline. At the level of the heart the anterior and posterior cardinal veins of the same side of the body become confluent in the common cardinal veins and turn ventrad to enter the sinus venosus.

Chicks of 4 days show little change in the. relationships of the cardinal veins. Later in development the proximal ends of the anterior cardinals become connected by the formation. of a new transverse vessel and empty together into the right atrium of the heart. Their distal portions remain in the adult as the principal afferent vessels *(internal jugular veins)* of the cephalic region.

The posterior cardinals lie at first in the angle between the somites and the lateral mesoderm. When the mesonephroi develop from the intermediate mesoderm, the posterior cardinal veins lie just dorsal to them throughout their length. Situated ventrally in the mesonephroi are small irregular vessels roughly paralleling the posterior cardinals and anastomosing freely with them. These are the subcardinal veins. They are arelic of the renal portal circulation of more primitive ancestral forms, and are of importance in the embryos of birds and mammals chiefly because of the way they are involved in the formation of the posterior vena cava. In young embryos the posterior cardinals are the main afferent vessels of the posterior part of the body. Later in development they are replaced in this function by the posterior vena cava. In 4-day chicks only the upper part of the posterior vena cava is indicated. It appears as a. slender vessel extending from the right subcardinal vein through a fold in the base of the dorsal mesentery to anastomose with venous channels in the liver. The formation of the vena .cava, in part as a new vessel, and in part by appropriation of already existing vessels, and the changes by which the posterior cardinals become reduced and broken up to form small vessels with new associations belong to stages of development beyond the scope of this book.

The Heart

The heart in adult vertebrates is a ventral, upaired structure. :s origin in the chick from paired primordia is correlated with the way the oung embryo lies spread out on the yolk surface. When the ventral wall is completed by the folding together of layers which formerly extended right and left over the yolk, the paired primordia of the heart are brought together in the midline. Their sequential fusion establishes the heart as an unpaired structure lying in the characteristic ventral position.

After the fusion of its paired primordia the heart is a nearly straight, double-walled tube. The primordial endocardium of the eart has the same structure and arises in the same manner as the endothelial falls of the primitive embryonic blood vessels with which it is directly continuous. The epipmyocardial laver of the heart is an outer investment which surrounds and reinforces the endocardial wall. As development progresses, the epi-myocardium becomes greatly thickened and is finally differentiated into two layers, a heavy muscular layer, the myocardium, and a thin nonmuscular covering layer, the epicardium.

In the apposition of the paired primordia of the heart to each other, the splanchnic mesoderm from either side of the body comes together dorsal and ventral to the heart. The double-layered supporting membranes thus formed are known as the dorsal mesocardium and the ventral mesocardium respectively. The ventral mesocardium disappears shortly after its formation, leaving the heart suspended in the body cavity by the dorsal mesocardium. Somewhat later the dorsal mesocardium also disappears except at the caudal end of the heart. Thus the heart comes to lie in the pericardial cavity unattached except at its two ends. The cephalic end of the heart remains fixed with reference to the body of the embryo where the ventral aorta lies embedded ventral to the floor of the pharynx, and the caudal end of the heart is fixed by the persistent portion of the dorsal mesocardium and the omphalomesenteric veins.

The straight tubular condition of. the heart persists but a short time. As the heart is lengthened, the unattached ventricular region becomes dilated and is bent out of the midline toward the embryo's right while the fixed outlet of the truncus arteriosus and the firmly anchored sinoatrial end are held in their original median position. This bending of the heart to form a U-shaped tube begins to be apparent in embryos of 30 hours and becomes rapidly more conspicuous, until by 40 hours the ventricular region of the heart lies well to the right of the embryo's body.

The bending of the heart to the side involves a considerable factor of "mechanical expediency." The initiation of the bending process depends on the fact that the heart is becoming elongated more rapidly than is the chamber in which it lies fixed by its two ends. The fact that the bending takes place to the side rather than dorsally or ventrally may be attributed to the impediment offered to its dorsal bending by the body of the embryo and to its ventral bending by the yolk.

The lateral bending of the heart attains its greatest extent at about 40 hours of incubation. At this stage torsion of the body of the embryo begins to change the mechanical limitations in the cardiac region. As the embryo comes to lie on its left side, the heart is no longer pressed against the yolk. As a result, the bent ventricular region begins to swing somewhat ventrad and lies less closely against the body of the embryo.

At about this stage of development the closed part of the U-shaped bend becomes twisted on itself to form a loop. In the formation of the loop the atrial region is forced slightly to the left (i.e., toward the yolk) and the truncus arteriosus is thrown across the atrial region by being bent to the right (i.e., away from the yolk) and then dorsocaudad. The ventricular region constituting the closed end of the loop swings dorsalward and toward the tail, possibly being crowded in this direction by the increasing flexion of the cephalic part of the embryo. Thus the original cephalo-caudal relations of the atrial and ventricular regions are reversed, and the atrial region which was at first caudal to the ventricle now lies cephalic to it as it does in the adult heart.

The atrial region and the ventricular region, which formerly were continuous without any linc of demarcation, are by this time beginning to be marked off from each other by a constriction. As both the atrium and the ventricle become enlarged, this constriction is accentuated . The constricted region is now termed the atrio-ventricular canal.

During the fourth day the truncus arteriosus becomes closely applied to the ventral surface of the atrium. As the atrium grows, it tends to expand on either side of the depression made in it by the pressure of the truncus. These lateral expansions are the first indication of the division of the atrium into right and left chambers which are later completely separated from each other. At the same time a slight longitudinal groove appears in the surface of the ventricle. This *interventricular groove* indicates the beginning of the separation of the ventricle into right and left chambers.

During the changes in the external shape of the heart which have been described, the whole heart has come to occupy a more caudal position with reference to other structures in the embryo. When the heart is first formed, it lies at the level of the rhombencephalon. As development progresses, it moves farther and farther caudad until at the end of the fourth day the tip of the ventricle lies at the level of the anterior appendage-buds.

During the third and fourth days of incubation, interesting changes become increasingly apparent in the ventricular portions of the cardiac wall. When the primordial tubular heart was first established, its endothelially lined lumen was smooth and fairly regular in diameter. Outside the endothelium, between it and the developing muscle, was a relatively thick layer of cardiac jelly. Beginning in embryos of 50 to 55 hours and rapidly becoming more marked, the myocardium shows irregular projections extending into the cardiac jelly. These projections are the start of the *trabeculae carneae* which are conspicuous features of the interior of the adult ventricles. As the trabeculae grow, the endothelial lining tends to extend between them and follow closely the configuration of the muscular strands separated from them only by a thin layer of cardiac jelly.

Once established, the trabeculation shows an increasingly richly branching pattern so that the ventricular wall becomes honeycombed by tortuous intertrabecular spaces, all of which are endothelially lined and all of which directly or indirectly communicate with the main lumen of the ventricle. This structural pattern is of great functional significance, for it brings the blood in close relationship to the growing cardiac muscle during the period before the coronary circulation to the myocardium has been formed. In human embryos the heart is beating and effectively propelling the blood through the embryonic vessels for about a month before its coronary vessels develop. During that crucial period the myocardium is entirely dependent on the blood that reaches it by way of the intertrabecular spaces.

Even in embryos of the fourth day the primordial epicardium consists of but a single layer of cuboidal cells. These are the forerunners of the mesothelium of the epicardium. Later in developmen a supporting layer of epicardial connective tissue will be formed between the mesothelium and the underlying myocardium.

Lying between the endocardium and the myocardium in the region of the atrio-ventricular canal and of the opening of the ventricle into the truncus arteriosus, there are loosely aggregated masses of cells which resemble mesenchymal cells in their loose, sprawling appearance. There is evidence indicating that at least some of these cells arise originally from the endothelium instead of all of them coming from the myocardium, as has been commonly believed. Whatever their source they move into the space between the two primordial layers of the heart using the cardiac jelly as a substratum on which to migrate. When there aggregated, they constitute what is called *endocardial cushion tissue.* This plastic connective tissue later takes part in the partitioning of the atrio-ventricular canal and in the formation of the connective tissue framework of the cardiac valves.

The definite establishment of the four-chambered condition characteristic of the adult heart belongs to later stages of development than those under consideration. By the end of the fourth day, however, the partitioning process is already begun. In reconstructions showing the interior of the heart the *interatrial septum* appears as a sickle-shaped partition cutting into the atrial lumen along the line where it is already narrowed by the pressure of the truncus arteriosus. On the interior of the ventricular wall, opposite the inter-ventricular groove, the trabeculae of growing muscle are especially abundant and project well into the ventricular lumen. Consolidation of these trabeculae establishes a definite septum which grows from the apex of the ventricle toward the atrio-ventricular canal. Convergent growth of the interatrial and inter-ventricular septa, and their ultimate fusion with a partition concurrently established in the atrioventricular canal, finally accomplish the division of the heart into right and left sides. At the same time truncus arteriosus is divided into pulmonary and aortic channels communicating, respectively, with the right and left ventricle. With the completion of these processes the heart is prepared to pump separate pulmonary and systemic blood streams.

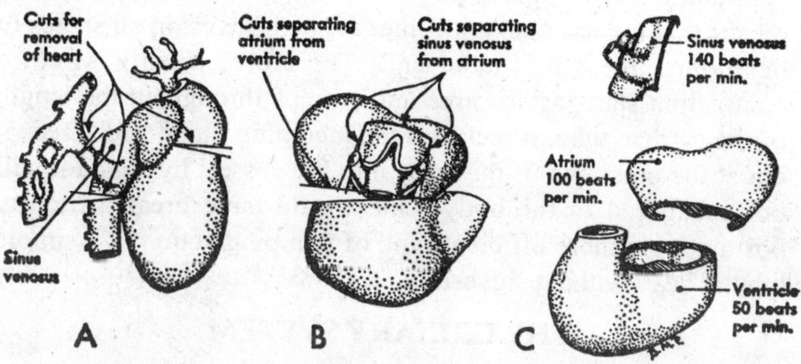

Figure 7.10 : Diagram showing the location of cuts made in the living heart of a 4-day chic embryo to demonstrate the relative pulsation rates of the myocardium from different regions. (A) Cuts for the removal of the heart. (B) Cuts to separate sinus, atrium, and ventricle. (C) The parts of the heart as isolated. The rates indicated are representative of the findings in such experiments rather than specific for any particular case.

There are exceedingly interesting physiologic differences in the myocardium of these basic cardiac regions. It will be recalled that, in dealing with the establishing of the cardiac tube, emphasis was laid on the sequential manner of its formation. Only as the foregut acquires a floor can the cardiac primordia meet and fuse in the midline. That means that the truncoventricular part of the heart is formed first, then fusion reaches the atrial region, and last of all the sinus venosus is established caudal to the atrium. Pulsation of the myocardium appears in these regions in the same sequence in which they are laid down. The first contractions appear in the ventricle before the atrium is fully established. The contraction rate of the ventricle is at first very slow. When the atrium begins to pulsate, the heart rate increases. We have already seen how transection experiments demonstrate that this increase in rate is due to the fact that the atrial myocardium has a faster inherent rate of pulsation and takes control of cardiac rhythm as a pacemaker.

If now we carry out similar transection experiments after the sinus venosus has been formed caudal to the atrium, we find its rate of contraction is higher than that of the atrium. There is thus a cephalo-caudal gradient in the intrinsic rate of myocardial contraction. The importance of this situation is far-reaching. It means that the pacemaking center of the heart is, in all stages of development, at its intake end. Therefore, when a wave of contraction starts at this area and sweeps throughout the length of the cardiac tube, it picks up the incoming blood and forces it out at the other end of the heart into the vessels by which it will be distributed to the body. One would have great difficulty postulating a more efficient type of pumping action in a simple tubular heart without valves.

THE URINARY SYSTEM

General Relationship of Pronephros, Mesonephros, and Metanephros

In the development of the urinary system of birds and mammals there are formed in succession three distinct excretory organs, pronephros, mesonephos, and metanephros. The pronephros is the most anterior of the three and he first to be formed. It is wholly vestigial, appearing only as a slurred-over ecapitulation of structural conditions which exist in the adults of the most primitive of teh vertebrate stock. The mesonephros of a bird or mammalian embryo is homologous with the adult excretory organs of fishes and amphibia. It makes its appearance somewhat later than the pronephros, and is formed caudal to it. The mesonephros is the principal organ of excretion during early embryonic life, but it also disappears in the adult, except for parts of its duct system which become associated with the reproductive organs. The metanephros is the most caudally located of the excretory organs and the last to appear. It becomes functional toward the end of embryonic life when the mesonephros is disappearing, and the persists permannetly as teh functional kidney of the adult.

Some of the main steps in the embryological history of the nephric organs, which it will be helpful to have in mind before taking up in detail any of the phases of their formation in the chick. The pronephros, mesonephros, and metanephros are all derived from the intermediate mesoderm, and are all composed of units which are tubular in nature. In the different nephroi these tubules vary in structural detail, but their functional significance is in all cases much the same. They are concerned in collecting waste materials from the capillary plexuses which are developed in connection with them. In the accompanying diagrams conventionalized tubules have been drawn to represent the three nephric organs. No pretense is made of representing either the exact shape or the actual number of the tubules.

In the first stage represented only the pronephros has been ablished. It consists of a group of tubules emptying into a common duct, led the *pronephric or primary nephric duct.* The primary nephric ducts either side are formed first at the level of the pronephric tubules and are n extended caudad, eventually reaching and opening into the cloaca.

As the primary nephric ducts are extended caudal to the level at which pronephric tubules are formed, they come in close proximity to the develop developing mesonephric tubules. In their growth the mesonephric tubules extend toward the primary nephric ducts and soon open into them. Meanwhile the pronephric tubules begin to degenerate. Thus the ducts which originally arose in connection with the pronephros are appropriated by the developing mesonephros. After the degeneration of the pronephric tubules these same ducts are called the mesonephric ducts because of their new associations.

At a considerably later stage outgrowths develop from the mesonephric ducts near their cloacal ends. These outgrowths form the ducts of the metanephroi. They grow cephalo-laterad and eventually connect with the third group of tubules developed from the intermediate mesoderm, the metanephric tubules. With the establishment of the metanephroi or permanent kidneys the mesonephroi begin to degenerate. The only parts of the mesonephric system to persist, except in vestigial form, are some of the ducts and tubules which, in the male, are appropriated by the testis as a duct systme.

The Pronephric Tubules of the Chick

The pronephros in the chick is represented by tubules which first appear at about 36 hours of incubation. The pronephric tubules arise from the intermediate mesoderm, or nephrotome lateral to the somites. They are paired, segmentally arranged structures, a tubule appearing on either side opposite each somite from the 5th to the 16th. Transverse sections passing through the 10th to 14th somites of an embryo of about 38 hours show the pronephric tubules favorably. Each tubule arises as a solid bud of cells organized from the intermediate mesoderm near its juction with the lateral mesoderm. At first the free ends of the buds grow dorsad, passing close to the posterior cardinal veins. Later the end of each tubule is bent caudad coming in contact with the tubule lying posterior to it. In this manner the distal ends of the tubules give rise to a continuous cord of cells, the primordium of the primary nephric duct. The pair of cell cords thus formed continue to extend caudad beyond the pronephric tubules and soon become hollowed out to form open ducts. When they eventually reach the level of the cloaca they turn ventrad and open into it.

The significance of the rudimentary structures in the chick, which represent pronephric tubules, can most readily be understood by comparing them with fully developed and functional pronephric tubules. The scheme of organization of a functional pronephric tubule. The ciliated *nephrostome* draws in fluid from the coelom. As the coelomic fluid passes in close proximity to the capillaries of a vascular ridge known as the *glomus,* waste materials from the blood are transferred to it. The nephric duct serves to collect and discharge the fluid passing through the tubules. In the pronephric tubules of the chick there are vestiges of a nephrostome opening into the coelom but the tubules never become completely patent and never acquire the vascular relations characteristic of the functional pronephros in primitive vertebrates. Shortly after their initial appearance the pronephric tubules begin to undergo regressive changes and by the end of the tourth day of incubation a few isolated epithelial vesicles are all that remain to chronicle the transitory appearance of the intermediate mesoderm caudal to the pronephros. The early steps in their formation are well shown in transverse sections of chicks of 29 to 30 somites (about 55 hours). In the posterior region conditions are less advanced than they are more anteriorly. Consequently by studying the posterior sections of a transverse series first and then progressing cephalad, a graded series of developmental stages may be obtained.

The Mesonephric Tubules

The mesonephric tubules appear first as cell clusters formed in the intermediate mesoderm ventro-mesial to the cord of cells which is the primordium of the primary nephric duct. The cells of the developing tubules acquire a more or less radial arrangment, and at the same time become more distinctly isolated from the surrounding mesodermal cells. By 55 hours of incubation the primordial cell cord representing the primary nephric duct has become hollowed out to establish a definite lumen. The most anterior of the mesonephric tubules also have acquired a lumen. Meanwhile the growth of the tubules has brought them in close association with the duct and in some of the more differentiated tubules indications can be made out of their opening into the duct. The more posterior mesonephric tubules do not become associated with the duct until somewhat later, but remain as a series of isolated vesicles.

The mesonephric tubule differs from the pronephric chiefly in its relation to the blood vessels associated with it. It develops a cup-like outgrowth into which a knot of capillaries is pushed. The cup-shaped outgrowth from the tubule is called the *glomerular capsule,* or *capsule of Bowman,* and the tuft of capillaries is termed *a glomerulus*. The diminutive form *glomerulus* is used to distinguish such small, localized capilary tufts from the continuous vascular ridge called a *glomus*. From the capillaries of the glomerulus there is free filtration of the fluid portion of the blood into the glomerular capsule. Then as this fluid passes along the lumen of the tubule there is selective resorption into the adjacent capillaries of such materials as salts and sugars, together with a large amount of water. It is this regulated resorption of substances in the nephric tubules that is responsible for the maintenance of the delicate salt-water balance which exists in the blood and tissue fluids. In this process waste materials are left within the tubule and carried by the balance of the water into the excretory duct and thence into the allantois for storage. The resorption of the great bulk of the water that originally entered the tubules which form no nephrostome but depend entirely on their glomerular apparatus for their fluid intake.

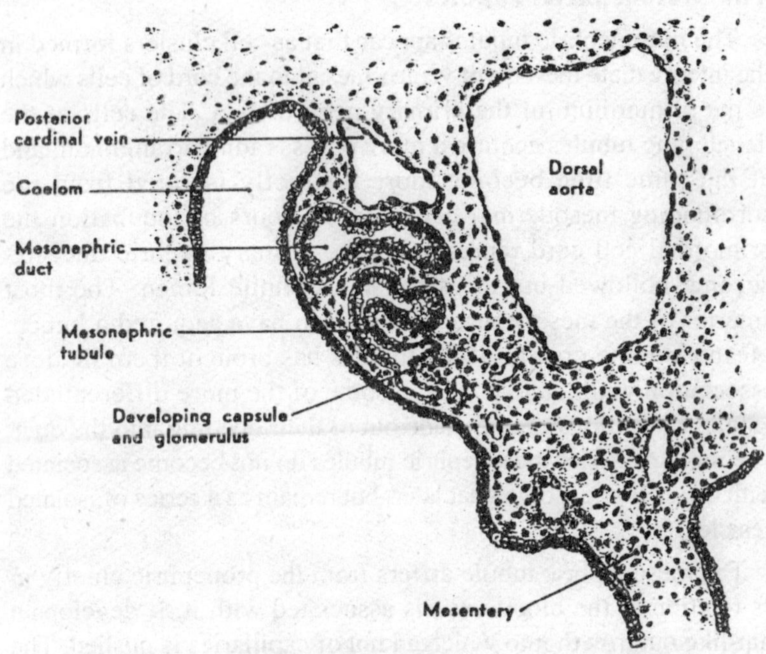

Figure 7.11 : Drawing from transverse section of 4-day chick to show mesephric tubule and duct.

In chicks of four days incubation the mesonephric tubules have not attained their full development, but it is possible to make out most of their fundamental parts. The tubules lying in the more cephalic part of the mesonephros have been longest established and are somewhat farther advanced in development than those lying more caudally in the mesonephros. Nearly all of the tubules, however, have become elongated and somewhat coiled. At one end they open into the mesonephric duct which acts as a main excretory channel. At their other end a cluster of closely packed cells, which is beginning to take on a cup-shaped arrangement, indicates the place at which the capsule will appear. The glomerulus is suggested by small vessels, or by endothelial sprouts, becoming organized in the concavity of the developing capsule. Once established, the glomeruli develop very rapidly. Circulation usually commences in them by the fifth day. From this time until about the eleventh day of incubation the functional activity of the mesonephros is at its height. After the eleventh day the developing metanephros begins to become active and the mesonephros degenerates. In discussing the distal portion of the allantois as a place of storage of waste materials, mention was made of the change in excretory product which occurs as development progresses. It will be recalled that urea is the predominant nitrogenous waste product of early stages and that uric acid is the chief form in later stages. The importance of this change lies in the low ,solubility of uric acid and the consequent lessening of the problem of its storage without toxic effects. This change in the character of the excretory products occurs as the metanephros takes over the excretory function from ,the regressing mesonephros.

The Metanephros

The differentiation of the metanephros and the development of the genital organs occur in stages which are too advanced to come within the scope of this book. Young mammalian embryos such as those of the pig or rabbit' make excellent laboratory material for the study of these more advanced phases of the development of the urogenital system and are widely used for this purpose. Those interested in studying this system as it develops in older bird embryos may consult a reference book such as Lillie's "The Development of the Chick."

In adult birds and mammals the body-cavity consists of three regions, pericardial, pleural, and peritoneal. The *pleural cavities* are paired, each of the pleural chambers being a laterally situated sac containing one of the lungs. The *pericardial chamber* containing the heart and the *peritoneal chamber* containing the viscera other than the lungs and heart are unpaired. These regions of the adult body-cavity are formed by the partitioning off of the primary body-cavity. or coelom of the embryo.

In the chick the coelom arises by a splitting of the lateral mesoderm on either side of the body. It is, therefore, primarily a paired cavity. Unlike the coelom of some of the more primitive vertebrates, the coelom of the chick never shows any indications of segmental pouches corresponding in arrangement with the somites. The right and left coelomic chambers extend cephalo-caudally without interruption through the entire lateral plates of mesoderm. This difference in the formation of the coelom does not imply any lack of homology between the coelom of the chick and that of more primitive forms. The process of coelom formation in the chick may be considered as being accelerated with a resultant slurring-over of the early phases. Or, to state the matter in another way, the coelom of birds and mammals first appears in a condition which is comparable with the coelom of more primitive forms at that period of differentiation when the segmentally arranged coelomic pouches have broken through into each other and their cavities have become confluent.

The coelomic chambers are not limited to the region in which the body of the embryo is developing. They extend on either side into the mesoderm, which, in common with the other germ layers, spreads out over the yolk surface. Large parts of the primitive coelomic chambers thus come to be extraembryonic in their associations. The portion of the coelom which gives rise to the embryonic body cavities is first marked off by the series of folds which separate the body of the embryo from the yolk. As the closure of the ventral body-wall progresses, the embryonic coelom becomes completely separate from the extra-embryonic. The delayed closure of the ventral body-wall in the yolk-stalk region results in the embryonic and extra-embryonic coelom retaining their open communication at this point for a considerable time after they have been completely separated elsewhere.

The same folding process which establishes the ventral body-wall completes the gut ventrally. Meanwhile the right and left coelomic chambers are expanded mesiad. As a result the newly closed gut comes to lie suspended between the two layers of splanchnic mesoderm which constitute the mesial walls of the right and left coelomic chambers. The double layer of splanchnic mesoderm which thus becomes apposed to the gut and supports it in the body-cavity is known as the *primary mesentery*. The part of the mesentery dorsal to the gut, suspending it from the dorsal body-wall, is the *dorsai mesentery,* and the part ventral to the gut, attaching it to the ventral body-wall, is the *ventral mesentery.*

When the dorsal and ventral mesenteries are first established, they constitute a complete membranous partition dividing the body-cavity into right and left halves. The primary dorsal mesentery persists in large part, but the ventral mesentery early disappears, bringing the right and left coelomic chambers into confluence ventral to the gut and establishing the unpaired condition of the body-cavity characteristic of the adult.

In considering the early development of the heart, the formation of the dorsal and ventral mesocardia was discussed. In their relation to the other mesenteries of the body, the mesocardia may be regarded as special regions of the ventral mesentery. In the most cephalic part of the body cavity, the gut lies embedded in the dorsal body-wall instead of being suspended by a dorsal mesentery as it is farther caudally. A ventral mesentery is, however, developed in the same manner anteriorly as it is posteriorly, andd when the heart is formed, it is suspended in the most anterior part of this ventral mesentery. The dorsal and ventral mesocardia may, therefore, be thought of as the parts of the primary ventral mesentery lying dorsal to the heart- and ventral to the heart, respectively.

When the ventral mesocardium, and a little later the dorsal mesocardium, break through, the primary right and left coelomic chambers. become confluent to form the pericardial region of the body-cavity. Later in development the ventral mesentery farther caudally disappears, so that caudally as well as cephalically an unpaired condition of the coelom is brought about

In the liver region a portion of the ventral mesentery remains intact. The liver arises as an outgrowth from the gut and in its development extends into this retained part of the ventral mesentery. That portion of the ventral mesentery which is dorsal to the liver persists as the *gastrohepatic omentum,* and that portion of the mesentery which is ventral to the liver persists as the *falciform ligament or ventral ligament* of the liver.

The entire dorsal mesentery persists and forms the supporting membranes of the digestive tube. In the adult its different regions are named according to the parts of the digestive tube with which they are associated, as, for example, mesogaster; that part of the dorsal mesentery which suspends the stomach; mesocolon, that part of the dorsal mesentery supporting the colon, etc.

The separation of the body-cavity into pericardial, pleural, and peritoneal chambers is accomplished by the formation of septa growing in from the body-wall. Consideration of the details of their formation would lead us into stages of development beyond the scope of this book. Those interested in following this or other phases of the later embryology of the chick will find in the bibliography references to more exhaustive books, and to some of the more recent original papers on its development.

8

Rate of Development

In wild birds the process of incubation provides potential stimulation of various kinds: sound vibration, also movement, changes in light and temperature. The embryo is active during the greater part of the incubation period, and although this activity is largely 'spontaneous', the embryo also interacts with its parents during the later, prehatching period.

We come now to the effect of different kinds of external stimulation on embryonic activity, development and the time of hatching. Certain signals have been found which result in the acceleration or retardation of hatching. These include non-vocal sounds, low frequency sound and also light.

THE SYNCHRONIZATION OF HATCHING IN QUAIL

Effects of stimulation on the time of hatching

When eggs of the bobwhite or Japanese quail are incubated in contact (as in a nest under natural conditions or closely packed in an incubator) they hatch within an hour or two of each other, whereas if they are separated by a few inches, hatching occurs over an extended period of time. Eggs in the same clutch can vary in a number of different ways which affect the incubation period, and it seems that consequent differences in developmental age can be cancelled out when the eggs are in contact. For this reason, it must be assumed that the embryos are capable of affecting each other.

Quail clutches not only have a smaller spread of hatching when the eggs are in contact, but the stimulation which brings about this effect must be of two kinds. When a few eggs are taken from a clutch and hatched in contact while the remainder are kept in isolation the contact eggs tend to hatch towards the middle of the hatching time of the isolates. This suggests that the most advanced eggs are held back by the retarded ones, while these more retarded embryos are brought on by the more advanced. It has been found possible to separate these effects. To show that hatching can be advanced a single egg was taken from each clutch as incubation began and then returned to the incubator 24 h later. These retarded

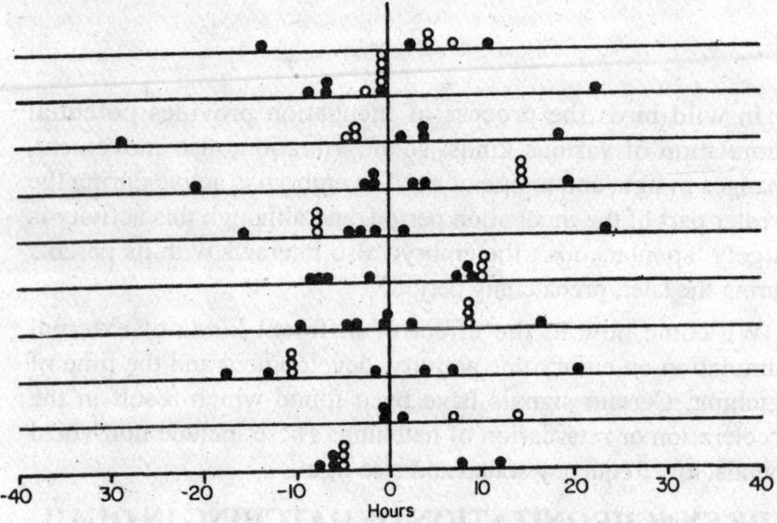

Figure 8.1 : The hatching times of eggs in contact (o) compared with isolated eggs in the same incubator at the same time (●). 0 hours = the mean hatching time of the isolates. Bobwhite quail, 10 clutches.

eggs were found to hatch at the same time as the eggs with which they were placed. At the same time, a younger embryo can be shown to affect an older one by delaying its hatching. In the bobwhite quail contact with a single egg given 24 h less incubation has been found to hold back the time of hatching. In this experiment each clutch was hatched with distances of at least 10 cm between the eggs except one egg, which was paired with another given 24 h less incubation, and this paired egg hatched in almost every case after the mean time of hatching of the unpaired isolates. Accelerating and retarding effects of mutual stimulation on the time of hatching in bobwhite eggs. Early attempts to repeat this experiment with Japanese quail failed; indeed, the delaying of hatching has so far always proved to be more *difficult,* than advancing it. Nevertheless, it has since been found possible to delay hatching in the Japanese quail by incubating an egg in a group with three or more retarded eggs. It would thus appear that more stimulation is needed to hold back the time of hatching in this species.

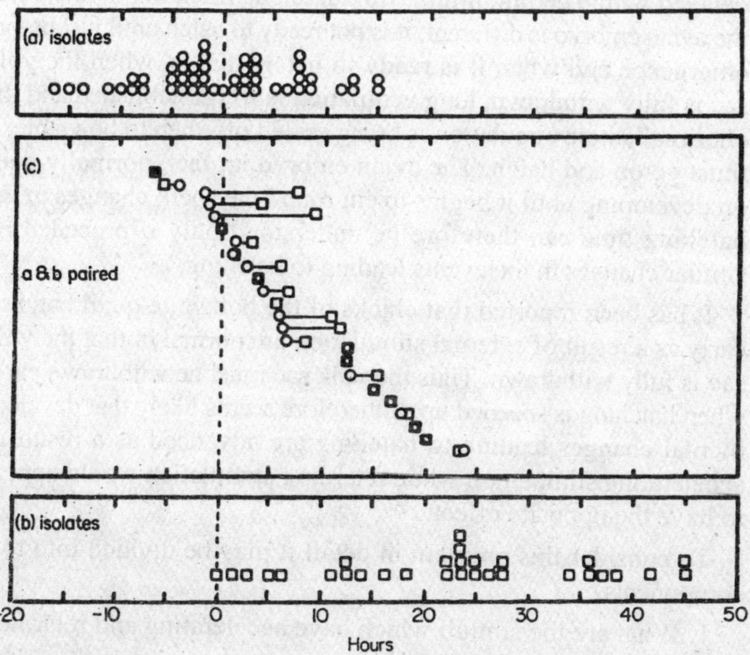

Figure 8.2 : Bobwhite quail. Accelerating and retarding effects of contact between eggs. (A) gives the hatching times of eggs put into the incubator first and (B) gives the hatching times of eggs where incubation was begun 24 hours later. (C) gives the hatching times of eggs incubated in pairs, with contact between the two. Within these pairs one egg was taken from group (A) and one from group (B). In some pairs the group (B) egg was accelerated and in some the group (A) egg was retarded. 0 hours represents the mean hatching time of the (A) isolates.

Pani *et al.* (1969a) have shown that a synchronizing effect occurs not only in natural clutches, but also in artificial clutches made up from eggs laid by different females. Eggs from early and late hatching strains of bobwhite quail hatched at the same time, when they were incubated in contact.

Problems arising from changes in hatching time

In the work described so far the hatching time has been changed by stimulation provided during the last 3 days, or even in the last 2 days before the normally expected time of hatching. An effect raises problems : hatching is not an event which can be triggered off in response to a signal. In many mammals the time of birth is not necessarily critical, for birth can be artificially advanced or delayed within certain limits without undue harm to the fetus. But the avian embryo is different: it is not ready to hatch until just before emergence and when it is ready to hatch, that is, when the yolk sac is fully withdrawn, lung ventilation is well established and the chorio-allantoic circulation is being sealed off, then to be viable, it must go on and hatch. The avian embryo, in fact, normally goes on developing until it begins to cut round the shell; changes in the hatching time can therefore be understood only if preceded by similar changes in the events leading to hatching.

It has been reported that chicks of the bobwhite quail hatched early as a result of external stimulation are normal in that the yolk sac is fully withdrawn. Thus the yolk sac must be withdrawn early when hatching is speeded up. It therefore seems likely that developmental changes leading to hatching are advanced as a result of accelerating stimulation while retarding stimulation would appear to have the opposite effect.

To consider this problem in detail it may be divided into two components

1. What are the stimuli which have accelerating and retarding effects on neighbouring embryos?

2. When, and on what developmental stages do they operate?

Signals which have accelerating or retarding effects on neighbouring eggs

Both audible sounds and also low frequencies just below the

auditory threshold have been recorded using a vibrator, on the shaft of which the egg rested, or a tripod constructed from three piezoelectric sound transducers on which the egg was supported. Using these techniques, it is possible to listen to and record sounds and vibrations produced by embryos during the period leading to hatching. (It is important, of course, in such an investigation, to use methods of recording which will not render the conditions of incubation abnormal : the embryo must be in a state to develop and hatch at the normal time. Similar methods have been developed by Kovach *et al.* (1990) and by Impekoven & Gold (1993).) Using one of these devices sounds can be heard from at least 6 days before the time of hatching in the fowl. At first the predominant signals are of very low pitch (from about 20-60 Hz). They continue until the time of hatching, and probably afterwards also. Each sound is usually short (not often more than 1 s, arid usually much less), they occur at irregular intervals and vary in rate of occurrence. Periods when they are frequent alternate with periods when they are absent, or rare. It seems very probable that they are produced for the greater part by the small, irregular, jerky, uncoordinated movements described by Hamburger (1984, 1988a, b) as occurring throughout the incubation period. When an egg from which the shell has been removed over the air space, is simultaneously recorded and observed, movements may be seen to be accompanied by these low-frequency sounds although in addition low-frequency sounds occur without movements being observed at the air space end of the egg. These, of course, could be produced by movements lower down within the shell. In an experiment where recorded sounds of this type were played back to an isolated egg, hatching was significantly delayed. It thus seems possible that bursts of low frequencies play a part as retarding signals, at a developmental stage when louder sounds of higher frequency are largely lacking.

A day or two before hatching in the fowl, and somewhat earlier in the quail, the predominant sounds change to include bursts of sharp, short signals of higher frequency. These occur at a rate of six or seven a second in quail, but three or four a second in the fowl and in bursts of one to about ten signals. They are produced by bouts of beak clapping. In the fowl beak clapping begins between the twelfth and sixteenth day, but the signals do not become audible even when amplified using the equipment described above until shortly before the onset of breathing. When the membranes are pierced they become louder. Using a pressure transducer to record movements of the embryo into and out of the air space, beak clapping signals appear not more than a few hours before the onset of breathing movements in the fowl (Vince & Tolhurst, unpublished). In the quail these signals, however, although quite loud, appear to have no effect on the time of hatching when recorded and replayed to an isolated egg. Beak clapping continues at intervals until the time of hatching.

The onset of pulmonary respiration introduces low-frequency pulses associated with breathing movements, although these are irregular and often intermittent at first. After an hour or two they become more regular at about one breath per second, but they remain relatively quiet for many hours (few signals on the sound channel). The possibility of the rather soft, low-frequency pulses associated with breathing having an effect on the time of hatching has been suggested. Experiments where signals are fed into isolated eggs at about the regular, early breathing rate (0·9-1·0 per second) have revealed a retarding effect on the time of hatching in the bobwhite, but not in the Japanese quail. In this case, however, the signals were artificial clicks of an amplitude higher than the clicks produced by another embryo, so that interpretation of these experiments is doubtful.

After embryos have been breathing regularly and fairly silently for several hours (the time varying with the species, see below) intermittent short click-like signals of higher amplitude begin to occur. These intermittent clicks then, quite suddenly, become more prominent, the respiration rate increases and regular loud clicking begins. These regular loud signals, which occur in all species and are produced by a special form of breathing appear almost certainly to act as accelerating stimuli in the quail. There are a number of different experiments supporting this view: the onset of clicking has been shown to be roughly synchronized in different eggs in a quail clutch, in addition, quail eggs stimulated by artificial clicks at about the normal click rate hatch earlier than unstimulated controls. These artificial clicks can be provided as auditory stimuli through a loudspeaker placed a few centimetres from the egg, or predominantly as vibration, by feeding them into a vibrator on which the egg rests. When the rate at which the clicks occur is varied from about 12 60 per second, an accelerating effect has been obtained in every egg tested, although an optimum appeared at about 3-6 clicks per second, when the amount of acceleration tended to be greater than at higher or lower rates.

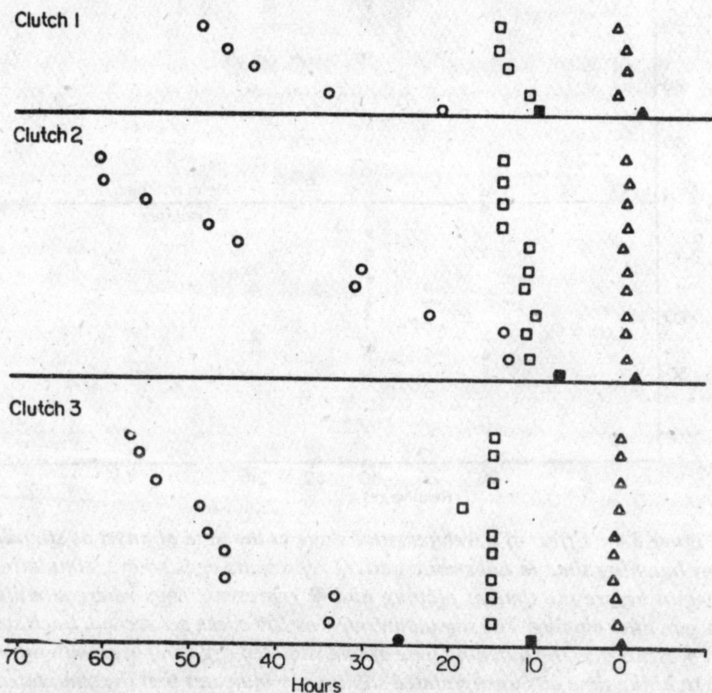

Figure 8.3 : Effect of contact between eggs on the rate of development leading up to hatching. Three clutches of the bobwhite quail. Within each clutch one egg was put into the incubator 24 hours late. Within each clutch symbols on one line represents one egg. p shows the time of pipping, o the time of onset of clicking and p the time of hatching. ● ■ ▲ indicate the retarded egg. In clutch 2 this egg pipped and began clicking at the same time.

Although it seems almost certain that clicking acts as the accelerating stimulus under natural conditions, it is known that other external stimuli (such as light) can have this effect also. It has also been found that changes in certain parameters can change a small retarding effect into acceleration. For example, whereas in the Japanese quail rapid clicks (above 80/s) delay hatching this result is obtained in the bobwhite only when the stimulation begins before the time of pipping. When eggs are already pipped the same signals advance hatching. A similar effect was obtained with click rates below 1¼ per second.

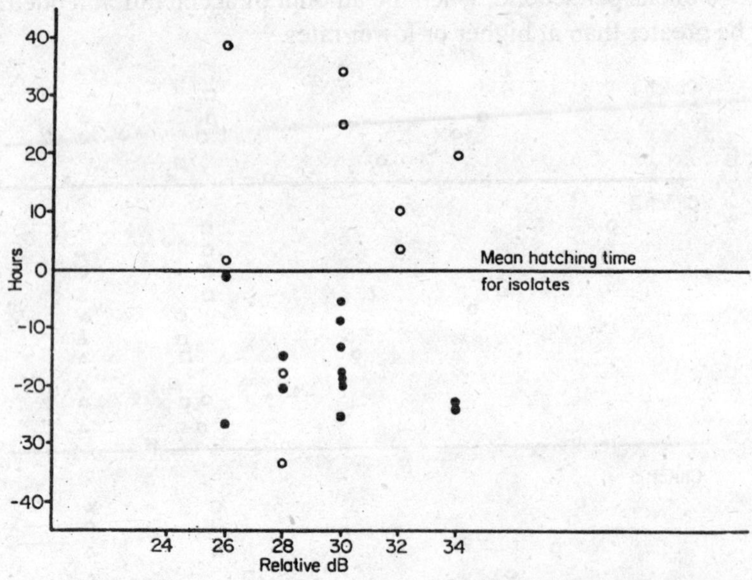

Figure 8.4 : Effect of developmental stage at the time of onset os stimulatrin on hatching time in bobswhite qail. O represents eggs where stimulatin was begun before the time of pipping and ● represents eggs where stimulation began after pipping. The stimulatin rate was 100 clicks per second. Each symbol represents also the hatching time of one stiulated egg compared with the mean hatchning time oits unstimulated sliblings. + indicates that the stimulated egg hatched late, and — indictes that it hatched early.

Much more work is needed in this area, but it would seem that, under natural conditions, a number of different signals could have an accelerating effect on hatching; for experimental purposes, however, regular loud artificial clicks at a rate of about three per second invariably have this effect.

The signals of greatest amplitude are produced at the time of hatching, and in a sequence which is very characteristic for all the precocial species considered here. These also may influence the time of hatching. Within a clutch the more retarded chicks occasionally emerge with a small part of the yolk not withdrawn into the body cavity, while the normal slight fall in the respiration rate which precedes hatching, does not always occur in accelerated quail eggs. Similarly when one egg is placed with a clutch 24 h late, it hatches at the same time as the clutch, but often after a very slightly shorter than normal period of clicking. Again, but only in the Japanese quail, it has been shown that the duration of clicking is slightly shorter in eggs hatched in contact with two given more incubation, than when in contact with others given the same amount of incubation. For such reasons it seems possible that signals, such as the sounds and vibrations produced by hatching or the rapid, soft vocalizations which slightly precede or accompany this activity, might trigger off the hatching activities in neighbouring eggs, at least in those embryos which are ready, or almost ready, to emerge. This possibility has not yet been tested systematically.

Effects of accelerating and retarding signals on the rate of development

Effects of natural stimulation. We know that the acceleration of hatching must result from pre-hatching development being advanced. To some extent this can be observed in eggs incubated in contact with others given more incubation. In such an experiment a single egg was taken from a clutch and put into the incubator 24 h late. Towards the time of hatching each egg in the clutch was examined at fourhourly intervals and the times of pipping, of the beginning of clicking and the time of hatching were noted for each egg. The time of pipping was found to be very variable. The beginning of clicking, however, was less variable, all eggs began clicking within a few hours of each other and in the eggs which pipped very late (such as the retarded egg) pipping was followed unusually quickly by the onset of clicking. The time of hatching was synchronized with more precision than the onset of clicking, but the duration of clicking, although affected, was very little shortened by contact between the embryos. From this, it was concluded that the onset of regular clicking is one developmental feature which can be advanced by accelerating stimulation.

These findings were extended using more precise recording equipment; three eggs were incubated in contact, one of which was recorded using a pressure transducer to transmit breathing movements. The recorded egg had been given (a) the same amount of incubation as the other two, (b) 24 h less incubation or (c) 24 h more incubation. The duration of the period of silent breathing which precedes clicking, the period of breathing preceding hatching which was accompanied by a click and also the respiration rate could be compared in all three groups. In the bobwhite quail the duration of clicking was roughly equal in all three groups although the period of silent breathing varied, being shorter in the eggs kept with two given more incubation and longer in those kept with two eggs given less incubation. It seemed, therefore, that the onset of clicking was advanced in the accelerated eggs and delayed in the retarded ones. In the Japanese quail the results were slightly different, in that the duration of clicking as well as the duration of the period of silent breathing were slightly shorter in the accelerated eggs, the other differences being negligible. So possibly both clicking and also hatching occurred earlier than would be expected in the accelerated Japanese quail. The stimulation had no effect on maximum respiration rates.

As a result of this experiment with very small groups (no more than three eggs) it was suggested that contact alone (regardless of the incubation ages of the eggs) could have an accelerating effect on development in the Japanese quail. This indicates a disadvantage in working with natural stimulation; there is more than one effect acting between embryos of different incubation age so that the more advanced can bring on the later ones while the latter may tend to hold back development in the more advanced. For this reason artificial stimulation provides a less ambiguous method of looking into possible effects on development.

Effects of artificial stimulation. Using artificial clicks at a rate of three per second, and stimulating eggs of the bobwhite quail, the report of *Pani et al.* (1989b) that the yolk sac is withdrawn early in accelerated embryos has been confirmed. At the same time it has been demonstrated that the allantoic circulation is sealed off early. This work also demonstrates a visible advancement in the bobwhite embryo when given 24 h of stimulation beginning 3 days before hatching. Furthermore it has been shown that the hatching muscle enlarges early in embryos of the domestic. fowl. (The hatching muscle has been shown by Fisher to develop to its full size at the time of pipping in the fowl, duck, and a number of other species.) Developmental stages earlier than this may also be affected. For example, Japanese quail stimulated by artificial clicks begin breathing early, although, so far, a number of attempts have failed to produce any effect on the onset of breathing in the bobwhite quail. When the onset of breathing is advanced in the Japanese quail, so also is lung aeration showing that the embryo is breathing air, and the movements recorded are not the earlier, low amplitude respiratory movements observed in avian embryos before the onset of pulmonary respiration and often before the embryo has taken up the hatching position.

True breathing, with aerated lungs, normally indicates that the embryo has attained the hatching position with the beak pointing upwards against, or more likely penetrating, the air space membrane (Plate 2.2 for the domestic fowl) (that it can occur without the attainment of this position in an embryo with the beak in the small end of the shell, or in one which has pipped the shell below the membrane need not concern us here). This hatching position is found in all unstimulated and accelerated embryos which have begun to breathe. What it does *not* necessarily indicate is that this position has been attained early; Hamburger & Oppenheim (1987) have found that embryos spend some time in the 'draped' position, while the membrane is slowly worn thin by beak and head movements. Whether or not the movements which bring the embryo into this position can be advanced by accelerating stimulation is not yet known.

To sum up : it would appear that the early hatching which results from accelerating stimulation is only one aspect of a changed developmental rate. The developmental sequence is normal, but more rapid. Different parts of the sequence, however, are not affected equally. The rate of development in an accelerated embryo appears to vary from the normal one at some, but not necessarily all developmental stages.

EFFECTS OF ACCELERATING AND RETARDING STIMULATION IN DOMESTICATED SPECIES

Quail species synchronize their hatching rather precisely; in an incubator a wave of hatching will spread across an egg tray within an hour or so including all viable eggs. This type of synchronization does not occur in other domesticated species, although in these species the same accelerating and (to a lesser extent) retarding processes appear to operate with a smaller effect. For example, McCoshen & Thompson (1998a) have shown that eggs of the domestic fowl, incubated in contact, have a smaller spread of hatching than others incubated in isolation. In confirming this finding it has been found that eggs in contact hatch with a similar mean incubation period, but with a smaller spread of hatching than isolates. This suggests that there could be a small retarding effect in the fowl, as well as a marked accelerating effect. In a similar experiment with ducks, eggs in contact were found to hatch with a smaller spread of hatching and in advance of isolates. It appears, therefore, that there is no retarding effect associated with contact between domestic ducks, although in the mallard Lien (1982) has reported both acceleration and retardation as an effect of contact with older or younger embryos. In the Leghorn chicken Lien found only acceleration.

Given artificial stimulation (clicks at the accelerating rate) eggs of the domestic fowl hatched earlier than unstimulated controls. As a result of artificial stimulation of the same kind, changes in the developmental rate have been observed in the fowl, the duck and the goose. In all three species the stimulated embryo breathed earlier than isolates, breathed for a shorter time before hatching, and also hatched earlier.

EFFECTS OF LIGHT IN THE FOWL

Behavioural responses to light are discussed in Section 6.8.4.

Under artificial conditions of incubation light has also been found to advance the time of hatching.

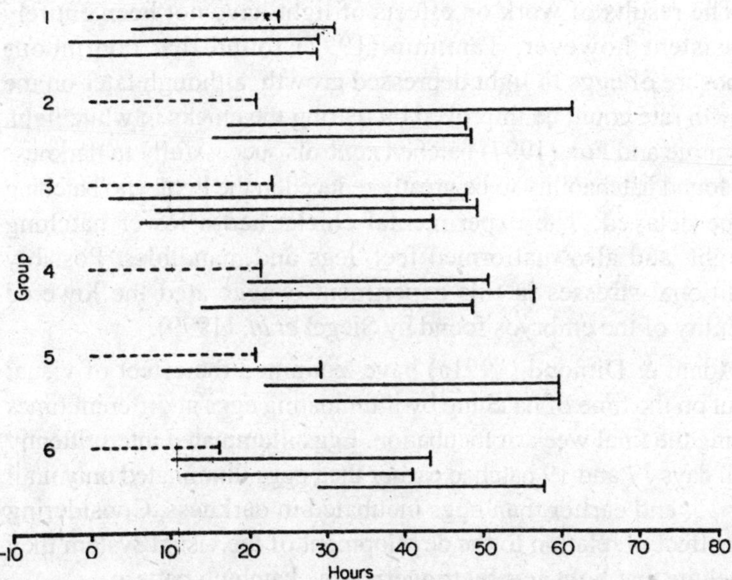

Figure 8.4 : Time of onset of breathing, duration of breathing and hatching time in six groups of eggs of the domestic fowl. Within each group one egg (dotted line) was given accelerating stimulation. Unstimulated eggs represented by continuous lines. Beginning of line represents the time of onset of breathing and the vertical line represents the time of hatching. Each group is arranged so that 0 hours indicates the time when the stimulated egg began breathing.

An effect of continuous light on the rate of development and time of hatching was reported in 1982 by Shutze and his colleagues and confirmed by Lauber & Shutze (1984). It was found that illumination of the eggs throughout incubation advances hatching by about 20 h. Hatching is advanced also, but to a lesser extent, when the incubator is illuminated for a single week, the first week giving a slightly larger effect than the second or third. No deleterious post-hatching effects were found in this experiment, and there were no effects on hatchability.

Siegel *et al.* (1989) found that light has the effect of accelerating early development as determined by the number of somites. Similar lighting treatment did not have any effects, however, if confined to the final week of incubation. When the incubator was illuminated throughout incubation the total incubation period was reduced as much as 30 h, without any significant deleterious effects apart from a 3 % reduction in hatchability.

The results of work on effects of light have not been entirely consistent however. Tamimie (1997) found that continuous exposure of eggs to light depressed growth, although later on the growth rate could be improved by rearing the chicks in white light. Tamimie and Fox (1997) hatched controls successfully in darkness but found hatchability to be greatly reduced in the light, and hatching to be delayed. The experimental chicks had a lower hatching weight, and also malformed feet, legs and mandibles. Possibly additional stresses in this experiment exaggerated the lowered viability of the embryos found by Siegel *et al.* (1999).

Adam & Dimond (1991a) have examined the effect of visual input on the time of hatching by illuminating eggs at different times during the final week of incubation. Eggs illuminated intermittently until days 17 and 19 hatched earlier than eggs illuminated only until day 15, and earlier than eggs incubated in darkness. Considering this effect in relation to the development of the visual system they conclude that light acts by triggering the hatching pattern.

Teratogenic effects of light on very early domestic fowl embryos (1-2 days of age) have been found by Kallen & Rudeberg (1984).

Light stimulation provided during the incubation period can, in addition, have an effect on the post-hatching behaviour of the chick. Dimond (1988), Adam & Dimond (1991b) and Dimond & Adam (1992) have tested chicks incubated in intermittent light for approach behaviour to a black and white rotating disc. Chicks stimulated until 19 days of incubation were found to be less fearful in this situation than dark-incubated chicks; they approached the stimulus more readily and gave fewer distress calls.

EMBRYONIC INTERACTIONS IN WILD SPECIES

The synchronization of hatching has been studied in the rhea by Faust (1990) and by Brunning (1993). The male in this species incubates eggs laid by a number of females, and he begins incubating before the clutch is complete. Both Faust and Brunning have noticed a possible difference of 12 days in the incubation period of individual eggs. According to Brunning isolated eggs require a 40-day period, while eggs in contact hatch after about 36-38 days. If, however, an egg is incubated in contact with more advanced eggs it can hatch after 28 days. This shortening of the incubation period is attributed to the transmission of calls and movements within the clutch.

In the yellow-wattled lapwing the five-egg clutch is incubated from the laying of the first egg, probably as a form of protection from the sun, yet hatching of all the chicks occurs within 24 h. In ducks, where incubation begins only after the clutch has been completed, there area number of reports of a very short spread of hatching, confirming the view that duck embryos also stimulate each other in such a way as to accelerate the development of retarded eggs. Leopold (1981) found a spread of hatching of between 4 and 6 h in the wood duck, Bjarvall (1997) found this to be between 3 and 8 h in the mallard and Fisher (1996) found the spread of hatching to vary between 8 and 16 h in the bluewinged teal, and 122 h and 132 h in the pintail. In the redhead he found that the spread of hatching increased from 10-60 h during the course of the breeding season, raising the possibility that interactions between embryos are affected by their viability. The tendency of duck embryos to synchronize their hatching is imputed by Hess (1992) to interactions between embryos and the incubating parent.

In the black-tailed godwit Lind (1981) compared the pip-hatch interval in different eggs. Within the same clutch he found that this interval became shorter in eggs which pipped later, suggesting that they may have been stimulated by the hatching activities of their more advanced siblings.

Eggs of all species examined click before hatching, raising the possibility that there is some accelerating effect in all species which incubate more than one egg. This would seem not to be the case, however. For example in Franklin's gull Fisher (1982) found an average spread of hatching of 24·3 h for two-egg clutches, and 46 h for three-egg clutches. Similar figures have been given for the blackheaded gull by Kirkman (1981). In some large birds such as owls, the eggs are laid at about two-, rather than the more usual one-day intervals and incubation begins with the laying of the first egg. The young of the Ural owl hatch at the same intervals and vary greatly in size. Under these conditions it seems likely that the developmental stages of different embryos within a clutch are too disparate for younger embryos to catch up.

THE EMBRYONIC RESPONSE TO STIMULATION

We have seen that incubation is an active process, the embryo is surrounded by potential stimulation of a number of different kinds. The embryo is active for the greater part of the incubation period, although this activity is considered to be largely spontaneous. That embryos of certain species, such as ducks and gulls, respond to parental calls. They respond in different ways (by calling, or by moving) and although their response is most marked during the last few days of incubation, in gulls at least it occurs also earlier than this.

The present chapter has also been concerned with the embryo's response to external stimulation, although this has been assessed in terms of changes in hatching time, or effects on developmental rate earlier than hatching. We need now to consider what mediates this effect. In reviewing this work Oppenheim (1993) puts forward the suggestion that both acceleration and retardation could be metabolic effects. They could result from sensory input (clicks, or low-frequency sounds) causing an increase or a decrease in the production of thyroxine.

That similar effects can be obtained by the use of hormones has been demonstrated by Balaban & Hill (1991). They found that injections of thyroxine at 16 or 17 days in the domestic fowl embryo result in an early onset of breathing, a slightly higher respiration rate and early hatching, while injections of thiourea have the opposite effect. Oppenheim (1993) has found no effect on the time of hatching when injections of thyroxine are given after the time of pipping, suggesting that several hours must elapse before such an effect can be obtained, and an hormonally controlled metabolic change is, indeed, likely to be observable only after a certain delay.

We do not yet know enough about the immediate effects of accelerating stimulation to decide whether these are mainly behavioural, mainly metabolic, or indeed whether such stimulation triggers both types of effect. So far there is no evidence that accelerating stimulation (such as clicks) can have the immediate type of behavioural effect shown in the gull by Impekoven & Gold (1993). Here it was found that embryonic motility increased in response to maternal calls as early as the fifteenth day of incubation. This increase was evident when levels of activity during a halfminute stimulation period were compared with levels occurring in preceding and succeeding minutes without stimulation.

The activity observed by Impekoven and Gold (1993) was of the random, jerky type. It seems possible that if changes in hatching time depend, at least in part, on a behavioural response, this response will be of the coordinated type leading to the hatching position. Preliminary work (Vince, unpublished) in the domestic fowl has suggested that somewhat more delayed effects, but on these co-ordinated behaviour patterns, can be obtained as a result of accelerating stimulation (clicks at a rate of 3 per second). In an experiment where the embryo was observed through a window in the shell, activity in a 15-min period of stimulation was compared with the activity of the same embryo during the preceding 15-min period of no stimulation and with the succeeding 15-min period of no stimulation. Under these conditions there was a significant tendency for the respiration rate to rise when stimulation began but to fall again before the end of the stimulation period. At earlier stages, before the onset of pulmonary respiration, and at stages when the embryo could be expected to make co-ordinated 'tucking' movements, there was a significant tendency for embryos to make more 'tucking' movements during periods of stimulation, than during the preceding no-stimulation period. Once they had begun, however (and they did not usually begin until almost the end of the 15-min stimulation period), these movements tended to continue, and thus were almost as frequent during the second period without stimulation. The co-ordinated movements occur in bouts, however, and not all embryos responded in this way. Only occasionally did a single co-ordinated movement occur soon after the beginning of stimulation.

It is not clear whether this rather slow response, resulting in the begin ning of a long bout of movements, can be considered as a purely behavioural response or not. It appears to be at a high level of complexity as these bouts are believed to depend on the temporary inhibition of the random type of embryonic motility and on neural organization at higher levels than the random motility.

TABLE 8.1 : VARIATION AND LENGTH OF INCUBATION IN NORMAL AND DEVOCALIZED PEKIN DUCKLINGS

			Length of incubation			
			Mean	*Standard deviation*	*Range (d & h)*	
Groups of ducklings	*n*	*Hatched* (%)	(d & h)	(h)	*Earliest*	*Latest*
Unstimulated vocal communal (eggs in contact)	274	76	26:23	7·8	25:12	28:00
Unstimulated devocalisolated	238	72	26:07	6·8	25:14·5	26:21·5

As we do not know what triggers these bouts under normal conditions, the problem of how the embryo responds to accelerating stimulation should, perhaps, be widened to include factors involved in the periodic shift to co-ordinated activity. A lot of work is needed here. In the meantime a few findings could bear on this problem. In the quail it has been found that, although the respiration rate rises more rapidly in stimu lated than in unstimulated embryos, it reaches about the same maximum rate before hatching. It has been shown that accelerating stimulation beginning at 18 days of incubation in the fowl is followed (in comparison with unstimulated controls) by a rapid increase in the size of the hatching muscle (suggesting the action of hormones), but no increase in embryonic size as indicated by leg length. There is also a curious finding that if accelerating stimulation is switched off after the onset of clicking the accelerated quail embryo appears to revert to an earlier stage similar to the onset of breathing, and tends to hatch late. The accelerating effect is thus not triggered in a simple way.

Recent evidence (Gottlieb, personal communication) adds support to the view that accelerated development can have a behavioural basis. His data compare the hatching time of normally incubated ducks (with the eggs in contact) with that of isolated eggs, where the embryo has been devocalized. The latter hatched earlier, and with a smaller spread of hatching. The devocalization procedure depends on removing the shell over the air space, pulling the embryo's head out of the shell, and thus artificially puncturing the allantoic and- inner shell membranes. Gottlieb considers that in this way the behaviours of the embryos may have been synchronized by placing them at the same stage of development.

9

Newly Hatched Bird

Previous chapters have shown that, as Hinde (1970) points out, hatching is not a zero-point for the development of behaviour'. Rather, the chances of successful hatching are maximized by the embryo's activity during the last few days of incubation. As lung ventilation is established before hatching, and as, for many species conditions within the nest are very similar in the pre- and post-hatching periods, it seems possible that there is more continuity during the trans-hatching period for the avian embryo and chick, than there is in mammals during the corresponding period (for continuity between the human fetus and neonate). Areas where continuity has been observed, and areas where new behavioural features have been demonstrated will be considered very briefly below.

First, the distinction between altricial and precocial species should be made clear. The early development of species which are, in varying degrees, altricial or precocial has been described by Nice (1983, 1992). Young birds which (i) feed themselves, are covered with down, have their eyes open and leave the nest during the day or two following hatching, are classed as precocial but to different degrees : (a) Megapodes are totally independent of their parents, *(b)* ducks and many shorebirds follow their parents but find their own food and (c) species like quail and fowl follow their parents but are shown food. Young birds which (ii) are fed by their parents include a special precocial group which leave the nest and follow their parents (grebes and rails) then, more dependent (a) semi-precocials which stay at the nest after hatching, although they are able to walk (gulls and terns), *(b)* semi-altricials which, although covered by down are unable to leave the nest (herons and hawks, eyes open after hatching) and those with eyes closed such as owls. Finally, (c) most dependent, are the true altricials, which have closed eyes, very little or no down, and are unable to leave the nest for some time like, for example, passerines.

Differences in the timing of developmental features in altricial and precocial species have been mentioned. These differences are very much in evidence after hatching. There are descriptions of altricial and precocial development in a few species. For example Nice (1983) has given a full description of the altricial song sparrow, while the altricial blackbird has been considered by Messmer & Messmer (1986) and the great tit by Gibb (1994). Kruijt (1994) has given a full description of posthatching development in the precocial Burmese red junglefowl.

REQUIREMENTS OF THE NEONATE

Warming and cooling

In most species parental brooding behaviour continues without much change during this transition stage between the incubation and the nestling periods (the blackheaded gull). In this way conditions suitable for warming or cooling the newly hatched bird are maintained at least until it dries out. In altricial species parental brooding behaviour continues, if only intermittently, until the nestlings gain control over their body temperature. Precocial chicks call loudly and in a distinctive fashion (the 'distress' call) when they are cold, or find themselves isolated and such calls stimulate parental brooding behaviour. This call is the same as that which may be heard from an egg which is cooling and in fact the alternation between this call and the 'pleasure' call, given by the embryo in response to warmth and gentle turning of the egg, also occurs in newly hatched chicks in response to warmth and contact.

Megapodes provide an important exception to this general picture of care which continues into the post-hatching period.

According to Frith (1982) the Mallee fowl's parental behaviour ends, in the case of the female with the laying of the egg, and in the male with his work of tending the incubation mound. The newly hatched Mallee fowl chick makes its way to the surface unaided, except indirectly by a loosening of the sand resulting from the male's building activities. (If a chick appears on the surface while the male is working it is scratched out of the way with the building material.) Once it has reached the surface it may rest in the sun for about an hour, and then run off into the scrub.

Righting and standing

Kovach (1990) has considered a possible developmental continuity between the co-ordinated and stereotyped movements which bring the embryo into the position for hatching, and contribute to cutting round the shell before emergence, and postnatal righting reflexes. In newly hatched domestic fowl he observed extension in both legs when the chick lies on its back with the head and neck stretched straight out; when the head is bent to the left, the left leg remains extended and the right leg is flexed, while the opposite occurs when the head is bent to the right. These postural reflexes are fully developed in the 19-day-old embryo and form a part of the co-ordinated and stereotyped movements from that stage onwards.

After righting itself the chick first creeps on its tarsi and later stands by extending its legs. The time for a chick to get up on to its feet and walk can be affected by its rate of development before hatching; artificially accelerated chicks took a little longer to stand up than unstimulated chicks. Similarly, abnormal prehatching conditions can have slightly damaging effects after hatching. In the bobwhite quail, lengthening the time spent in the shell before hatching lengthens also the time required for uncurling the toes after emergence (equivalent effects in the human neonate).

As in righting and standing, other reflex activities have become functional during the incubation period, although the extent to which they function during that time is not yet understood, in all cases. Detailed trans-hatching studies are needed here.

Social attachments in young birds

In precocial species the parents and young leave the nest within the first few days (or even within a few hours after hatching) and move round together, with intermittent bouts of feeding and brooding. Here there is an obvious need for a mechanism which will keep the group together. The concept of 'imprinting' (although it arose in the context of species recognition) has led to the analysis of factors involved in the 'following' response in young precocial birds. Many years ago it was found that a newly hatched fowl, duckling or gosling will follow almost any moving object in much the same way as it will follow its parent. Spalding (1893) considered this behaviour to be an instinctive response to visual and auditory stimulation : 'chickens as soon as they are able to walk will follow any moving object. And, when guided by sight alone, they seem to have no more disposition to follow a hen, than to follow a duck, or a human being ... there is the instinct to follow and ... their ear, prior to experience, attaches them to the right object'.

In addition to parameters related to the formation and maintenance of this parent-chick bond recent work has considered the part played in its establishment by pre-hatching experience and activity. Tschanz (1988) and Impekoven (1991) have demonstrated that *embryos* of the guillemot and gull respond to parental calls in such a way as to query the last part of Spalding's statement: in fact neonates may well have become attached to their parents' calls while still in the egg. Further, Gottlieb has shown that the duckling's preference for the calls of its own species has been facilitated by hearing calls from neighbouring eggs during the last few days of incubation, and also, by hearing its own calls after it begins to breathe about 3 days before hatching.

Feeding behaviour

Given the right temperature conditions, a full complement of reflex responses and the necessary social adjustments, the newly hatched bird's more basic requirements are completed by food. Much of the feeding situation is obviously a new development, but continuity with pre-hatching activities is still discernible even apart from swallowing movements which began half-way through incubation. In the smaller altricial species food is an immediate and urgent need if the nestling is to grow quickly enough to be viable and here food is supplied by the parent. Food is brought to the nest either fresh or partially digested. In response to mechanical stimuli (the parent landing on the nest) or parental vocalizations, the nestling 'begs' for the food, by stretching up its neck vertically, with the beak wide open and with typical 'begging' calls. This situation has been analysed in nestling thrushes by Tinbergen & Kuenen (1989). The 'begging' posture in the great tit nestling is shown in Figure; in this species the vertical feeding position is exaggerated, presumably owing to the exceptionally deep nest cup. Although it is weak in the first day or two, the begging pattern can consist of the extension of the limbs, neck and body, raising the possibility that it could be derived from hatching movements if, as has been suggested, these are the same in altricial as precocial species.

Pecking for food in the domestic fowl has been much studied. Spalding (1873) observed chicks with no previous visual experience and noted that their pecking was accurate, although they were less successful in seizing the object between the mandibles at the moment of striking. Subsequently many attempts have been made to separate this skill into innate and acquired factors. In the meantime Kuo (1982d) observed head, beak and swallowing movements in the domestic fowl embryo and came to the conclusion that pecking in the chick arises from elements which have already matured and been exercised in a co-ordinated fashion in the normal course of embryonic development. This view, however, does not take into account the perceptual problem where new and important factors are introduced and which will be discussed below.

The transition from incubation to feeding of the young in the ring dove has been analysed by Lehrman (1995). The squabs are fed 'crop milk' regurgitated by the parent. Lehrman found that one of the factors leading to regurgitation is tactile stimulation of the crop; this occurs when the newly hatched squab first lifts its head unsteadily and moves it against the parent (i.e. a 'begging' movement from the position in the nest in which the squab has hatched). Thus he found that a successful first feeding results partly from the parent's hormonal state and partly from the presence and activities of the young, both arising in the normal course of incubation.

In the semi-precocial herring gull, Tinbergen (1993) has given a detailed account of the stimulation eliciting feeding behaviour. Gull chicks beg for food by pecking at the red spot on the underside of the parent's bill and their activities result in the regurgitation of food by the parent. In a very detailed study of pecking in the laughing gull Hailman (1997) has analysed the type of stimulation eliciting pecking in this species and the form and accuracy of their pecking movements. This brings us again to problems beyond the scope of embryonic life; the perception of objects, colour, distance and visually guided movements in space.

Oiling and preening

Care of the plumage is. important for both precocial and altricial birds. In precocial species preening begins very early; Kruijt (1984) mentions preening movements on the first day after hatching in the Burmese red junglefowl, in particular pecking and nibbling movements, although restricted to a small area of the plumage, occur at that time. The possibility that these movements could be derived from beak clapping in the embryo, needs to be considered. In the altricial song sparrow and redstart Nice (1983) reports that the first preening movements occur before there are any feathers to be preened, and they begin about 5 days after hatching.

In ducklings, which may be led to water on the first or second day after hatching, oiling of the down is an essential factor in preventing their drowning. The way in which this occurs stresses the importance of continuity between the conditions of incubation and those in the post-hatching period. Many species do not have functional preen glands at hatching, and the ducklings' down becomes oiled by the mother's plumage as she broods them.

WHAT IS NEW IN THE LIFE OF THE NEONATE?

The precocial embryo may well have had some experience of

light and shade, may well have responded to auditory, tactile and proprioceptive stimulation and will have exercised its muscles for a considerable time, during the last few days of incubation. Nevertheless it will be faced with new conditions and new problems from the time of hatching. Of these, movement in space, patterned vision and presumably also a wide range of tactile stimuli seem likely to be among the more important.

In precocial and semi-precocial species new co-ordinations appear after hatching; pecking (although it may have pre-hatching antecedents) at the red spot on the parent's bill is a clear example of this. Perception of distance appears to be good, but not perfect; for example Hailman (1987) finds that the laughing gull chick, although it will haltingly approach a specific stimulus such as food, may not always position itself at the optimum distance for striking in the first trials. Response to depth, as in the 'visual cliff' experiment varies from one species to another, the difference appearing to depend on the nesting habits of the species. For example, Kear (1987) has demonstrated a difference between newly hatched ground-nesting and holenesting ducks in that ground-nesting ducklings avoid moving over the 'precipice' whereas the hole-nesting ducklings (which have to jump to the ground soon after hatching) do not. Wehrlin & Tschanz (1999) have demonstrated that the ledge-reared guillemot chicks are able to perceive a precipice optically and avoid it, and the same response occurs in the cliff nesting kittiwake. However, comparing results from two surface- and one cliff nesting species Hailman (1998) suggests that chicks of all gull species are hatched with an ability to perceive and avoid a cliff-edge.

In such cases centres of co-ordination must have been already formed in the nervous system, although they could not have functioned before hatching. Colour, again, must be a new experience; it has been found, however, in certain species (such as ducks) that the neonate pecks more frequently and more vigorously at certain colours, such as green, while domestic chicks respond more readily to orange and blue.

Taylor *et al.* (1997) and Sluckin *et al.* (1996) have been able to demonstrate the existence of a tactile discrimination in the domestic chick. This, associated with warmth, must play a part in making brooding effective. Although tactile sensitivity occurs early there would seem to be a possibility of change in the type of stimulation received cutaneously after the chick has dried. Sluckin *et al.* and Taylor *et al.* found chicks to prefer a smooth to a rough surface.

Movement in the chick has been mentioned already in areas where it could rather obviously be derived from pre-hatching patterns of behaviour; walking, running, hopping on different types of substrate (or indeed on the ground *at all)* involves the development of new, albeit very rapidly developed skills. An example of this is given by Kruijt (1994) in the prococial junglefowl a day or two after hatching. Here, again, the muscular co-ordinations will probably have been exercised before hatching, but the relationship with the ground, balance, and the stimuli which call forth this exquisitely polished performance must be new.

EFFECTS OF NEW ENVIRONMENTAL STIMULI ON NEURAL MECHANISMS

In dark-incubated domestic fowl Bateson & Wainwright (1972) have shown that a 30-min previous exposure to light has a dramatic effect on the rate at which a chick acquires a preference for a visual stimulus, in the imprinting situation. More recently, Bateson & Seaburne-May (1993) have found that the first approach to a flashing light occurs more rapidly the longer the previous exposure to the light. These findings are consistent with those of Dimond (1998) and Adam & Dimond, showing a con tinuity across hatching. Bateson and Seaburne-May suggest that activation of the visual pathways simply by use (exposure to light) may mean that visual stimuli become more effective in evoking behaviour thereafter. This effect is specific, as after previous exposure to sound chicks were found to be less responsive to the flashing light. (Possibly this specificity could explain Hunt's (1989) work on prenatal conditioning to sound; this was found to be effective postnatally only until the chick was given experience of light, food, water, etc.)

Thus, although there is obvious continuity across hatching, there is also change. There is evidence that the functioning of higher levels of the nervous system matures rather rapidly after hatching, and there is evidence also that this follows from the enormous increase in stimulation which occurs at this time. Growing evidence that exposure to stimulation can affect the connectivity of neurons in the central nervous system should help to elucidate the nature of the change which comes after hatching.

Figure 9.1 : Hopping in the chick of the Burmese red junglefowl.

CONCLUSIONS

Considering the hatched chick in the light of previous chapters there is evidence of behavioural, as there is of physiological continuity. This aspect of development has been stressed by Kuo (1982a, b, c, d, e, 1983, 1997) and Schnierla (1995), who have both been interested in embryonic antecedents of chick behaviour.

Although there is obvious continuity some of Kuo's examples of it are now known to be unlikely (for example, that active movements can be Stimulated by heart beats in the first few days of incubation, and also that activity can be conditioned at an early age. Even so, the idea that pecking in the chick arises on the basis of separate head, beak and neck movements all of which have been exercised *in ovo* still seems unexceptionable. In addition Kovach (1990) has pointed to continuity in righting movements across hatching and Corner *et al.* (1993a) have observed a continuation of embryonic stereotyped and co-ordinated movements in the younger sleeping chick. Suggestions could be made of further continuities but the difficulty is to *demonstrate* their existence. New and refined techniques could be needed here.

The problem of techniques applies not only to continuities across hatching. Although qualitative accounts have been given, a detailed analysis of embryonic movements has long been thought desirable and it could be valuable to consider continuities (and also possible discontinuities) from the beginning of active movements until well after hatching. This could show when particular behaviour patterns appear and disappear, it could take into account the direction and amplitude of movements, the elements included in the random, jerky motility and the question whether any of these are included in the stereotyped and co-ordinated movement patterns, as well as providing a basis for assessing new features in the days after emergence from the shell.

Together with the gradual development of sensory systems, of responsiveness and the effects of greatly increased sensory input after hatching, understanding of motor development is needed before we can assess the continuity between embryo and chick or say how much of chick behaviour is new.

10

Hormones in Development

In the course of elucidating the endocrine status of an organ it is usual to remove it surgically from the animal and then, if obvious changes result, to attempt replacement therapy either by means of grafting or by injecting simple extracts of the organ. If this course of action succeeds then the active principle or principles will be isolated and identified and their precise function determined by further experimentation. One difficulty in this approach is that the amount of substance required to elicit a response is not likely to be known, and the observed response might therefore not be the normal response. The use of specific inhibitors overcomes to a large degree these problems and, as we shall see below, the results of experiments using these substances can produce more satisfactory results than those using the purified hormones.

There is also another complication. The fact that a response may be elicited following the application of a particular stimulus which may mimic a normal response of the developing animal, does not necessarily mean that this is the normal stimulus.

The avian. egg, free from maternal influences, offers the endocrinologist an excellent experimental system, yet its potential has, surprisingly, not been fully exploited and large areas remain to be explored. Moreover, while it has been known for many years that certain endocrine glands become functional during incubation, it is only relatively recently that the extent of hormonal influences in development has been appreciated.

THYROID HORMONES

Development of the Thyroid Glands

The thyroid of the domestic fowl arises as a midventral diverticulum of the pharynx on the second or third day of incubation. By the fifth day the anlage has become two-lobed and the cells have migrated towards the third aortic arch as double sheets surrounded by connective tissue. These sheets separate into clusters of cells and colloid droplets form intracellularly by the eighth day. The droplets coalesce in the clump of cells to produce first two large extracellular droplets then a single follicular space. The follicles increase in size by the fusion of adjacent follicles and by the division of the surrounding cells. The apical surface of the cells is characterized by cilia. The Golgi bodies are well developed by the eighth day and rough endoplasmic reticulum increases from about the eleventh day to achieve the appearance of the adult thyroid by the fourteenth day.

Synthesis and Secretion of Hormones

There is little evidence of diversity in the biosynthetic mechanisms of the thyroid hormones in the different classes of animals and it seems likely that the avian embryo conforms to this pattern. Iodide ions are oxidized to iodine by a peroxidase system and the iodine reacts with the phenolic ring of free tyrosine or of tyrosine in proteins to form monoand diiodotyrosines. Thyroglobulin is synthesized as polypeptide subunits containing uniodinated amino acids, which are then modified by the addition of substituted hexosamines to form a glycoprotein. Iodination occurs from the completion of the glycoprotein subunit until the molecule of thyroglobulin is secreted into the follicular colloid. Thyroxine is then synthesized from one free iodotyrosine molecule and one iodotyrosine bound into the thyroglobulin molecule. The thyroglobulin molecules thus contain thyroxine, triiodothyronine, mono- and diiodotyrosine. The thyroxine and triidothyronine are liberated by the hydrolysis of the thyroglobulin – the other amino acids are re-utilized by the thyroid cells.

In mammals triiodothyronine is between five and seven times more potent than thyroxine but in the hatched bird at least, and probably the embryo, these hormones have a similar potency. This is due to similar degrees of binding by plasma protein.

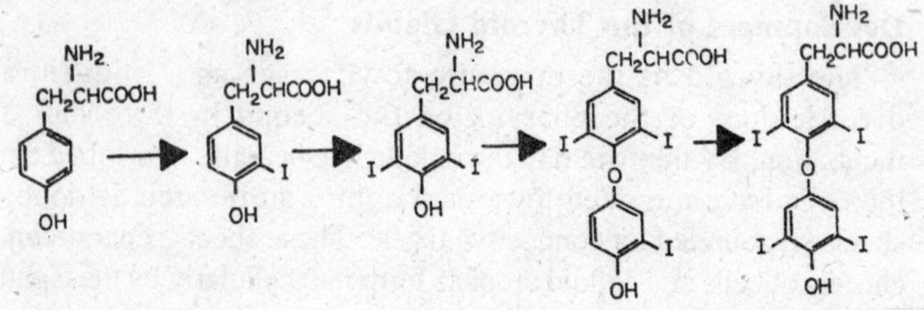

Figure 10.1 : Metabolic pathway for the synthesis of the thyroid hormones.

The age at which iodine is concentrated by the thyroid has been variously estimated as 51 days, 7 days, 8 days, 9 days (Hunt, 1983), 10 days and 11 days. Since monoiodotyrosine can be detected in the thyroid by 82 days, diiodotyrosine by the ninth day and thyroxine by the tenth day it seems likely that the concentration of iodine begin no later than the eighth day. It is generally agreed, however, that secretion of thyroxine and presumably, triiodothyronine, begins on the tenth day, Trunnell & Brayer (1983) having shown that this is the time when the thyroid itself matures.

Secretion of the thyroid hormones (it is not known in what proportions thyroxine and triiodothyronine appear in the blood of the embryo) is controlled by the hypophyseal hormone, thyroid-stimulating hormone (thyrotrophin). The development of the thyroid and its ability to concentrate iodine are influenced by vitamin B_{12}, insulin and also by the bursa of Fabricius.

It is probable that the rate of secretion of the hormones is low until the fifteenth day. There is a marked increase in the protein bound iodine content of the blood during hatching.

Metabolic Effects of the Thyroid Hormones

Although the age at which secretion of the thyroid hormones begins is known with some precision, less is known of the fundamental actions of these hormones. Their role in controlling basal metabolic rate is generally accepted but they also appear to play a major part in promoting growth and differentiation during development. However, it is clear that in the case of the bird perhaps half of development *in ovo* occurs without the modifying influences of thyroxine and triiodothyronine. Indeed, while the tissues of the young embryo are probably sensitive to their action (Portet, 1990) there is evidence that even low concentrations - 100 ng per egg – are teratogenic.

The injection of thyroid hormones at about the time the thyroid begins its own secretory activity has been found to produce variable results and has thus failed to provide much information on the significance of the hormones in the second half of incubation. Romijn *et al.* (1982) found thyroxine stimulated metabolic rate but no increase in growth rate resulted. Portet (1990), on the other hand, could not detect any change in either parameter, while Beyer (1982) found similar increases in growth rate and oxygen consumption with overall increases in metabolic rate.

The increase in oxygen uptake, growth rate and the significant, though temporary fall in the hepatic stores of glycogen are all thought to result from the onset of thyroidal secretion. Attempts to delay the depletion of hepatic glycogen by treating the embryo with the goitrogen, thiourea, have not been completely successful though partial decapitation ('hypophysectomy') prevents completely this fall.

While exposure of the embryo to enhanced concentrations of the thyroid hormones has not produced consistent results, reduction or suppression of the natural secretion of the thyroid with substances such as thiouracil or thiourea or by hypophysectomy is generally agreed to result in a decreased growth rate, decreased metabolic rate, hypertrophy of the thyroid and a delay in the time of hatching though these effects do not become manifest before the tenth day when the thyroid normally begins active secretion. Glycogenesis is reduced in both liver and yolk sac membrane following treatment of the embryo with thiourea. Again these effects are not seen before the tenth day.

The thyroid hormones influence bone development. Adamson and Ingbar (1997a, b) have shown that triiodothyronine, but not thyroxine, stimulates the uptake of neutral amino acids by developing bone. Indeed there is some evidence to suggest that thyroxine may inhibit growth. The morphogenesis and keratinization of the scales of the leg are also influenced by the thyroid for thiourea delays development while exogenous thyroxine advances it. Similarly the development and pigmentation of the down feathers are affected by the thyroidal status of the developing chick.

The importance of the thyroid hormones in the morphogenesis and histogenesis of the duodenum is now well-established. Development of the duodenum ceases at 171 days in hypophysectomized embryos and is retarded in embryos treated with thiourea but continues if thyroxine is made available. The thyroid also apparently influences adrenal development, though it is uncertain whether this relationship is direct or whether it involves the pituitary.

Bargman & Gardner (1997) have implicated the thyroid gland in the development of the auditory system.

The Thyroid and Hatching

In the course of studies on hatching chicks of the domestic fowl Freeman (1962) noted that oxygen consumption began to rise about 2 h before the chick began the final breaking down of the shell. Indeed since there was no concomitant increase in overt activity it seemed possible that this rise was hormonally induced and that the hormone might therefore be considered to be the hatching stimulus. In view of their effects on metabolic activity, the thyroid hormones were soon considered for this role.

As we have already seen from the results of work using goitrogens, the thyroid is capable of influencing the rate of development and the time of hatching. However, this action may be seen as a general response by the tissues and it does not necessarily follow that the thyroid produces a specific stimulus which initiates active hatching. Treating the embryo at a relatively late stage of development with thiouracil was found to have little effect on development but a delaying action on the time of hatching. The relationship between thyroid hormones and the stimulation of active hatching was further emphasized by Freeman (1984) who showed that by treating the embryo with either thiouracil, triiodothyronine, thyroxine or thyrotrophin at 18+ to 19 days the time of hatching could be greatly influenced without, significantly, affecting the time at which pulmonary respiration was initiated. Unfortunately neither Balaban & Hill (1991) nor Oppenheim (1993) have been able to confirm that part of the work employing thiouracil. Both groups have found that the thyroid hormones accelerate hatching, though when the treatment was given at 17 days, pulmonary respiration was also significantly advanced suggesting that at this earlier time the embryonic response is general rather than specific.

Thus while the thyroid is implicated in stimulating active hatching specifically the hypothesis remains unproven. Protein-bound iodine concentrations in the plasma have been shown to double during the last 30 h of incubation, from 6·3 to 13-2 $\mu g\ I^{-}\ 100\ ml^{-1}$. However, further work on this aspect of thyroidal physiology is needed.

The thyroidal status of the embryo appears to affect the rate at which yolk material is absorbed by the embryo during the latter stages of incubation. Clearly this is important in relation to the movement of the yolk sac into the abdominal cavity prior to hatching. Finally it should be noted that the development of the hatching muscle *(musculus complexus)* may be influenced by the thyroid hormones.

PARATHYROID HORMONE AND CALCITONIN

It will be convenient to consider these two hormones together for though they are secreted by different glands - parathyroid hormone (parathormone) by the parathyroid glands, calcitonin by the ultimobranchial bodies - both are concerned with the control of calcium metabolism. Their functions both in the embryo and in the adult have not yet been fully elucidated, though it seems probable that they have antagonistic effects.

Development of the parathyroids and ultimobranchial bodies

The parathyroid glands probably arise from the endoderm from the third and fourth visceral pouches. The pair on each side generally fuse to give one glandular mass which is close to, or even attached to, the thyroid gland of that side. Accessory parathyroid tissue often occurs in the thymus and in the ultimobranchial bodies, while active ultimobranchial C-cells are often found in the parathyroids.

The ultimobranchial bodies are derived from the endoderm of the sixth visceral pouch. The cells migrate to a position below the parathyroids and thyroids. Details of their development may be found in Dudley (1982). Their structure and ultrastructure are described by Stoeckel & Porte (1997a, b; 1999a, b) and Hodges (1990). The endocrine cells of the ultimobranchial bodies are distinct and are termed C-cells.

Secretion of parathormone and cakitonin

Sun (1982b), on histological grounds, suggested that there was no secretion of parathormone during development *in ovo*. However, later work has generally indicated that synthesis of the

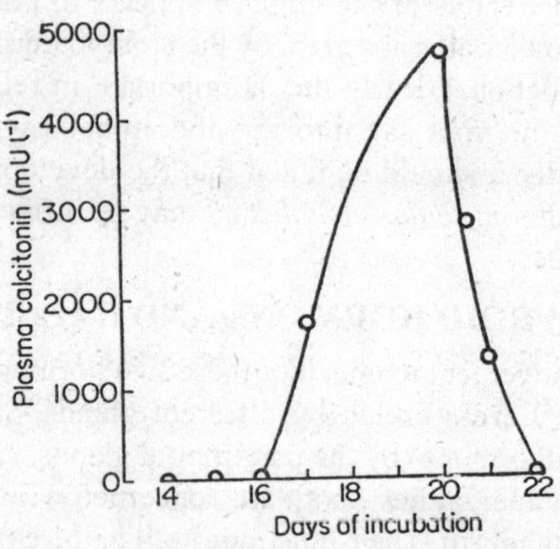

Figure 10.2 : Plasma calcitonin concentration during the development of the fowl.

hormone begins on about the eighth day and that there may be actual secretion from about 101 days. The rate of synthesis seems likely to increase sufficiently by the thirteenth day for storage vesicles to be formed in the cells. Bone resorption through parathormoneinduced osteoclastic activity can be stimulated by the eleventh day, though Jones (1990) believes the action of parathormone *in ovo* is to inhibit ossification rather than stimulate resorption.

Calcitonin contains about 32 amino acids and has a molecular weight of the order of 4500. Stoeckel & Porte (1999a) presented histological evidence of secretion from the eleventh day though direct measurements show no appreciable amounts in the plasma before the seventeenth day. The concentration of calcitonin in the ultimo is very high in the embryo 2500 mU kg^{-1} rising to 5200 mU kg^{-1} 3 days after hatching. It reduces thereafter to reach 600 mU kg^{-1} in the adult.

Significance in development

The role of these two hormones in calcium metabolism of the adult is still equivocal. The situation with regard to the embryo is even more problematical.

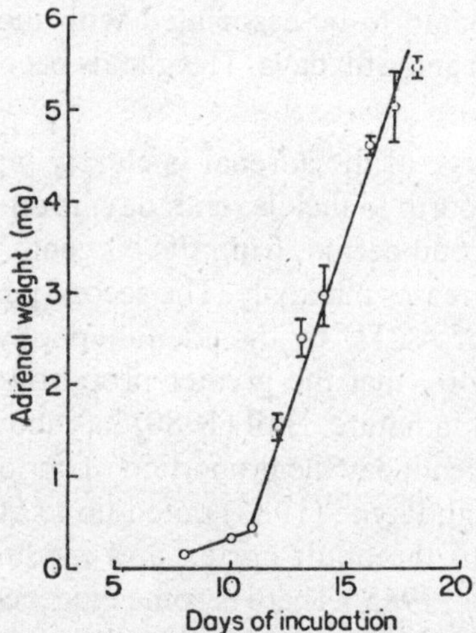

Figure 10.3 : Growth of the adrenal glands of the fowl during incubation. Each point is the mean net weight of the paired glands. Stand ard errors are indicated.

In the adult parathormone stimulates a rise in plasma calcium through bone resorption while calcitonin probably has the reverse effect. The importance of controlling plasma calcium levels within fine limits for the proper functioning of muscles and nerves cannot be over-emphasized. One might reasonably expect the parathyroids and ultimobranchial bodies to be active, particularly during the period of active calcium resorption from the shell, that is from 14-19 days. However, no evidence has been presented to indicate any parathyroid involvement in this process. On the other hand the excretion of phosphorus by the developing embryo may be partly controlled by parathormone.

The secretion of calcitonin is probably maximal immediately before hatching and coincides with a lowered plasma calcium concentration. The significance of this is unknown.

ADRENAL HORMONES

Development of the adrenal glands

The primordial adrenal cortical cells arise from the mesoderm medial to the mesonephros and can be distinguished by the fourth day in the domestic fowl. The cortical cells develop rapidly and by the seventh day they are arranged characteristically in cords. Medullary cells begin to be associated with these cortical cells between the fourth and fifth days. The glands become encapsulated by the tenth day.

The growth curve of the adrenal is clearly biphasic. The first period, from the fourth to the eleventh day, is one of slow growth but during the second period, from the eleventh day to hatching, the growth rate increases markedly. The second period is controlled by the secretion of ACTH by the adenohypophysis and it is not surprising, therefore, that the greater proportion of the adrenal should be cortical in nature. Hall (1980) has shown that from the thirteenth to eighteenth day the proportion of cortex increases from 65% to 77% though Payne (1985) noted up to 90% in the newly hatched chick. (In the adult cortex and medulla are in equal proportions - Payne, 1985.) There is some evidence that the growth rate of the adrenal of the male is greater than that of the female.

The ultrastructure of the developing adrenal is described by Straznicky *et al.* (1996) and Hall & Hughes (1990).

Synthesis and secretion of adrenal hormones

Adrenal steroids. The biosynthetic pathways have not been

elucidated completely in either the embryo or the adult, but they almost certainly follow the general scheme. The ability of the 14-day-old embryo to synthesize 11-deoxycorticosterone, corticosterone and aldosterone from acetate has been demonstrated by Bonhommet & Weniger (1997a, b) and semi-quantitated by Pedernera & Lantos (1993). It is possible that, as in the adult, the main secretory product is corticosterone. It is not certain whether hydrocortisone (cortisol) and cortisone. are synthesized by the bird, though the former steroid has been shown to have considerable activity in the embryo.

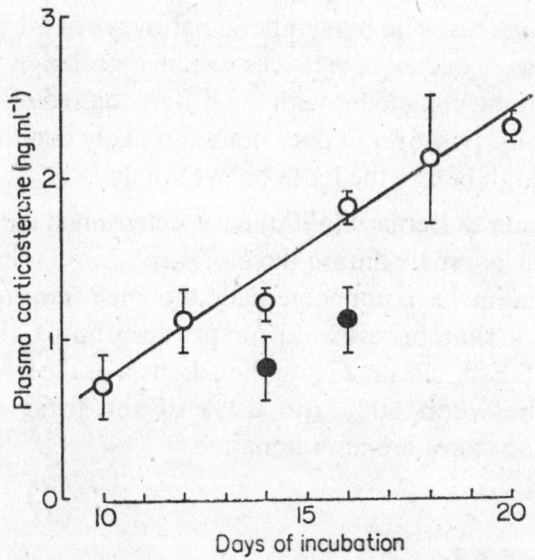

Figure 10.4 : Plasma corticosterone concentrations during development of normal (O) and hypophysectomized (●) fowl.

It has been shown both histochemically and by bioassay that the adrenal cortical cells are able to synthesize corticosteroids from the fifth day of incubation - only 24 h after their differentiation. However, natural secretion *in vivo* does not begin until the eighth day when the endogenous secretion of ACTH begins. The establishment of the pituitary-adrenal axis is thus achieved at an early stage of development. Early work had suggested that the adrenal cortex functions, at least to a degree, autonomously but this has now been disproved. Woods *et al.* (1991) found the axis established by 141 days, Adjovi & Idelman (1999) by the twelfth day but more sensitive techniques have shown it to be established as early as 8 days.

While the amount of circulating corticosterone rises progressively during development, Piddington (1990) and Woods *et al.* (1991) have found that the secretion rate undergoes a marked rise at about 141 days indicating an increase in the rate of turnover.

It is generally agreed that the hypothalamus is not involved in the control of the normal secretory pattern of the adrenal cortex, at least during incubation though is probably required in the mediation of the stress response.

Catecholamines. The biosynthetic pathways of adrenaline and noradrenaline. Catechol substances can be detected in the adrenal at 4 days - a time coincident with the differentiation of the adrenal medullary cells. However, it does not seem likely that the secretory rate is very high before the tenth or twelfth day.

Wassermann & Bernard (1990) have determined the adrenaline content of the adrenals during development. Similar estimates of the noradrenaline have not been made though a semi-quantitative study shows that noradrenaline predominates throughout development. Estimates made on the adrenals of the neonate show that from between 60% and 80% of the total medullary catecholamine stores are noradrenaline.

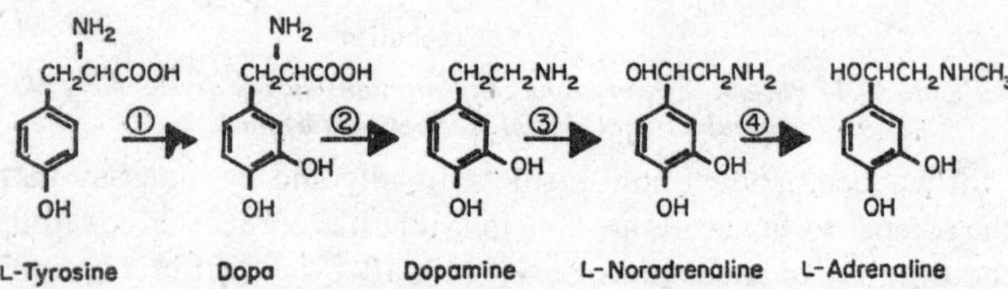

Figure 10.5 : Metabolic pathway for the synthesis of catecholamines. The enzymes involved are as follows:

(1) Tyrosine 3-hydroxylase

(2) Dopa decarboxylase

(3) Dopamine hydroxylase

(4) Phenylethanolamine N-methyltransferase.

Significance of adrenal cortical hormones in development

As we have already seen the developing fowl synthesizes at least two cortical hormones, corticosterone and aldosterone. While their roles in adult life are reasonably well understood, their function in the developmental processes remain, with one important exception, to be elucidated.

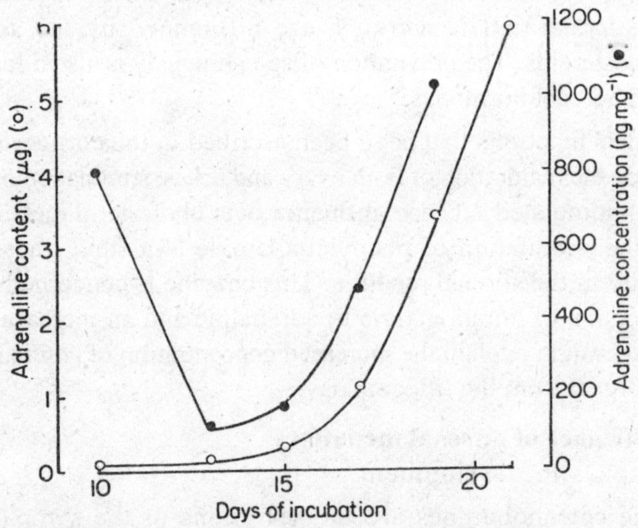

Figure 10.6 : Adrenaline concentration (l) and content (O) of the adrenaline during development of the fowl.

The exception concerns the differentiation and maturation of the duodenum in the last few days before hatching. Several authors believe that both the adrenal and the thyroid are concerned in this though Pedernera (1991) suggests that the adrenal alone is involved. The effects of corticosteroids on alkaline phosphatase activity may be relatively non-specific for there is indirect evidence that. they act on that enzyme in bone and thus influence bone formation. There is, furthermore, a suggestion that alkaline phosphatase may be stimulated by another, as yet unknown, mechanism for the activity of this enzyme is elevated in mature embryos treated with a corticosteroid inhibitor.

There is some evidence that the adrenal corticosteroids are involved in the maturation of the exocrine pancreas. Exogenous hydrocortisone stimulates the accumulation of amylase, procarboxypeptidase A, chymotrypsinogen and to a less extent endonuclease. This maturation of the pancreas is initiated on the fourteenth day, a time, as we have seen, when adrenal cortical activity is increasing. Hijmans & McCarty (1986), Dautlick & Strittmatter (1990) and Strittmatter (1992) have also found that the maturation of the small intestine, and the activities of the maltase and sucrase that it secretes, are influenced by the adrenal corticosteroids. The prevention of splenomegaly is also a function of the adrenal hormones.

Other functions that have been ascribed to the corticosteroids include the maturation of both ovary and testes, stimulation of (Na^+ + K^+)-stimulated ATPase, influcnce over cholesterol metabolism and the stimulation of phenyletholamine-N-methyl transferase activity in the adrenal medulla. This enzyme is concerned in the conversion of noradrenaline to adrenaline and an increase in its activity might explain the increased concentration of adrenaline in the adrenal from the fifteenth day.

Significance of adrenal medullary hormones in development

The catecholamines are the secretions of the sympathetic nervous system of which the adrenal medulla is but part. Thus this tissue differs from other endocrine tissue in being part of a larger functional unit which is under the direct control of the central nervous system. While the sympathetic nerve cells have long axons and secrete catecholamines locally at the surface of the cells they innervate, the medullary cells lack axons and secrete their hormones into the circulation thereby causing general reactions throughout the body. Because of this dual system it is necessarily difficult to subscribe with ccrtainty any catecholamine-induced reaction to the adrenal medulla. Indeed the role of catecholamines in the developmental processes generally remain largely speculative, though there has recently been speculation that they play an important role in early morphogenesis.

It is possible that the progressive increase in blood pressure during development may be due to circulating catecholamines for Girard (1997, 1993b) has shown that the circulatory system is sensitive to both adrenaline and noradrenaline from the sixth day.

Immediately prior to the initiation of hatching (i.e. before pulmonary respiration is established) there is a profound mobilization of glycogen stores of the embryo. This certainly involves the liver, probably the yolk sac membrane but, significantly, neither the heart nor the brain. The mechanism of this mobilization has been studied only in the liver. Adrenaline stimulates glycogenolysis though noradrenaline is ineffective. It seems possible, therefore, that the adrenal medulla is concerned though the data of Freeman & Manning (1991) show that, on a molar basis, glucagon is perhaps twenty times more potent than adrenaline in mobilizing glycogen. This might be thought to indicate that glucagon is the more likely stimulus. However, the natural mobilization of hepatic glycogen is impaired by pretreatment of the embryo with a- or E3-receptor blocking agents indicating a role for the catecholamines rather than glucagon. At the same time it has been shown, albeit in rabbits, that the mobilization of hepatic glycogen is most rapidly achieved by stimulation of the splanchnic nerves innervating the liver cells. Thus, paradoxically, while catecholamines may well be implicated, it seems more likely that they emanate from the autonomic nervous system rather than the adrenal medulla.

PANCREATIC HORMONES

In addition to the two well-known hormones, insulin and glucagon, the avian pancreas secretes a physiologically-active substance, at present known simply as avian, polypeptide. Its significance remains to be elucidated.

The role of insulin in carbohydrate metabolism in the adult remains problematical, indeed its significance has occasionally been questioned. However, evidence is now being collected to confirm its role in carbohydrate metabolism. As we shall see, its role in embryonic development has been partly elucidated.

Glucagon, a powerful glycogenolytic hormone in mammals and birds, is a particularly potent lipolytic agent in the latter. Recent research has demonstrated that a glucagon-like substance is also secreted by the small intestine. It seems likely that the pancreatic and gut glucagons have different roles in metabolism, though that of the gut factor has yet to be determined.

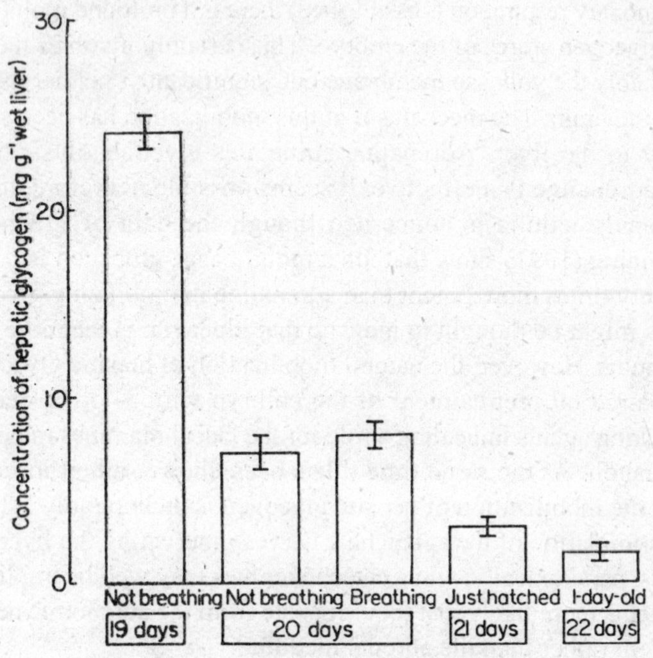

Figure 10.7 Hepatic glycogen stores during hatching. Standard errors are indicated.

Development of the endocrine pancreas

The pancreas arises as three separate primordia : one is a dorsal evagination of the gut while the other two are lateral evaginations. These anlagen arise, in the fowl, after three days and eventually fuse. The various cell-types of the islets of Langerhans begin to differentiate early. Their distribution is not random; there is a predominance in the splenic lobe which is derived from the dorsal anlage. Studies with the light microscope suggested α- and β-cell differentiation on about the eighth day but with the electron microscope these cell types can be recognized on the third day - that is shortly after evagination. A third cell type, the 8-cell, can also be differentiated by the tenth day. Its significance remains to be elucidated: some suggest it is either equivalent to the a-cell or is a precursor of the p-cell; others recognize it as a distinct cell type and suggest it is concerned with gastrin secretion.

Structure and secretion of insulin and glucagon

Insulin. Avian insulin, like its mammalian counterpart, is composed of two chains of amino acids. There are six different amino acids in the sequence of fowl insulin as compared with porcine and ox insulins; fowl and turkey insulin are identical but duck insulin has three different amino acids in the sequence as compared with fowl insulin.

Insulin is secreted by the p-cells, from the fourth or fifth day though not to any significant extent before the twelfth or thirteenth day. The importance of the splenic lobe as a source of insulin is shown by the data of Kedinger *et al.* (1992): 66% of the total pancreatic insulin is found in this tissue which itself represents 6% of the pancreatic tissue.

Attempts to measure the concentration of circulating insulin during development have been unsuccessful, probably because the concentrations are below the sensitivity of the methods currently available. At hatching the concentration is about 6 tLU ml^{-1}.

Glucagon. Two molecular species of pancreatic glucagon have been isolated from the duck, one with a molecular weight of 3000 the other of 6000. The jejunum and ileum, but not the duodenum, secrete only the 'large' molecular form. In the adult, however, the circulating molecule, whether from pancreas or gut, is the large form. No information regarding other avian species is available.

The pancreatic α-cells begin to secrete glucagon on the third, fifth, eighth or tenth day. There are, as yet, no data on circulating glucagon concentrations.

Role of insulin in development

Zwilling (1981) was the first to show that injected insulin causes hypoglycaemia and enhanced the concentration of glycogen in the yolk sac membrane. The latter observation was confirmed by Thommes & Mathew (1999) who further established that the response could be elicited as early as the fourth day of incubation. This response is probably mediated through UDPG-glycogen synthetase. The increase in the concentrations of glycogen in the, yolk sac membrane and liver during the third week of incubation may therefore be controlled by the natural secretion of insulin stimulating the activity of this enzyme.

Insulin also affects the maturation of the hepatocytes. Benzo & de la Haba (1992) have shown that insulin is needed for the development of smooth endoplasmic reticulum in the hepatocytes. This insulin, moreover, has to be complexed to zinc.

The lipogenic mechanisms of the heart tissues are stimulated by insulin, as is glucose uptake by the same tissue though only from the seventh day. It should be remembered, however, that the rate of insulin secretion is probably low until the third week.

Insulin has also been implicated in the stimulation of the uptake of iodine by the thyroid and of the uptake of neutral amino acids by the nervous system.

Role of glucagon in development

While glucagon plays a very important role in lipid metabolism of the hatched bird, being. a potent mobilizer of lipid from the adipose tissue, it does not seem that this activity is present during incubation or at least only to a very restricted degree. This reduced activity is due partly to the relative immaturity of the adenyl cyclase-cyclic AMP system. Glucagon has activity in the carbohydrate metabolism of the embryo and may well have an important role in controlling circulating glucose concentrations.

Glucagon stimulates glycogenolysis in both the yolk sac membrane and the liver. This is brought about, at least in the yolk sac membrane, by an activation of phosphorylase.

It has already been noted that there is a drastic mobilization of glycogen just prior to the onset of breathing. Given the glycaemic activity of glucagon it would seem possible that this hormone is responsible.

On balance, however, the evidence currently available suggests that this mobilization is mediated by the adrenal medulla or the autonomic nervous system, i.e. by catecholamines. Perhaps the most important piece of evidence for this interpretation is that a- and β-adrenergic receptor blockers impair the natural mobilization of glycogen.

During hatching the adenyl cyclase-cyclic AMP system of adipose tissue matures so that glucagon is able to stimulate both lipolysis and glycogenolysis. However., by reducing the dose rate to 1 $\mu g\ kg^{-1}$ Freeman & Manning (1991) found that only the lipolytic mechanism was stimulated suggesting that this becomes the more important role of glucagon after hatching.

Early reports that glucagon has growth-promoting activity have not been substantiated.

GONADAL HORMONES

Development of the genital system

Initially the genital system of the avian embryo develops similarly in both males and females - for about 5 days in the house sparrow, 7 days in the domestic, fowl and 71 days in the duck. Sexual dimorphism commences thereafter.

In the female the Miillerian ducts, derivatives of the urogenital ridges, and the gonadal tissues develop asymmetrically: the left ovary and duct develop normally while the right ovary and Miillerian duct develop indifferently at first and then degenerate. Both mesonephric (Wolffian) ducts degenerate though they usually persist into post-embryonic life.

Both gonads and Wolffian ducts develop in the male to give the testes and vasa deferentia, whereas the Miillerian ducts regress. In the duck and fowl regression begins on the ninth day and is complete by the fourteenth.

Details of the structure of the gonads may be found in Romanoff (1990). The ultrastructure of the ovary is described by Simone-Santoro (1997, 1999), Scheib (1990) and Budras & Preuss (1993) and of both ovary and testes by Dubois & Cuminge (1997) and Rahil & Narbaitz (1992).

Ovarian oestrogenic hormones are synthesized and secreted by medullary interstitial cells, for those cells have high concentrations of lipids and cholesterol and A^5-3phydroxysteroid dehydrogenase. Ovarian androgenic secretion may be a property of the cells of the primary albuginea and stroma.

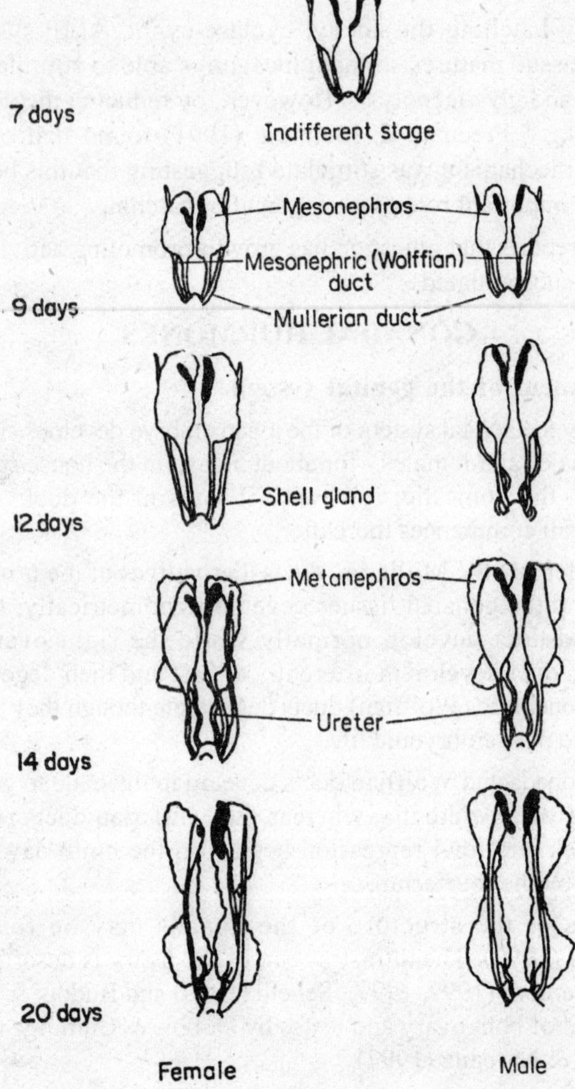

Figure 10.8 : Development of the male and female genital tracts of the fowl.

In the male early steroid secretion is found in the cord cells but is progressively lost during development. At about 8^1 days in the fowl (132 days in the Japanese quail) the interstitial cells of the differentiating testes show A^5-3p-hydroxysteroid dehydrogenase activity.

Synthesis and secretion of gonadal hormones

The general metabolic pathways of the biosynthesis of gonadal steroids in mammals: (the synthesis of pregnenolone). Studies on the bird have generally confirmed the similarities between the two classes.

Synthesis of hormones certainly begins shortly after the gonadal anlagen appear on the fourth day and there is some evidence that this begins earlier and is more active in the female. Between the fourth and sixth days, while the gonadal tissue is in the indifferent state, oestrogen and oestradiol-173 are synthesized though Galli & Wassermann (1993) suggest that these are synthesized only by gonads destined to become ovaries. On the seventh day, still in the indifferent state, the gonads synthesize testosterone.

As differentiation of the gonads proceeds the ovary produces progressively more oestrogens with oestradiol-173 always exceeding oestrone production. Oestriol production begins on the fourteenth day. Similar results have been obtained with the guinea fowl and duck. The testes do not produce significant amounts of oestrogens but increasingly more testosterone.

Initially gonadal hormone secretion is probably autonomous. The pituitary-gonadal axis is established from about the thirteenth day. Gonadotrophins have been found to increase markedly the secretion of oestradiol-17β in both sexes but have little effect on oestrone production. Akram *et al.* (1993), however, found that the secretion of oestrone and oestradiol-17β was unaltered following hypophysectomy.

Significance in development

That secretion of gonadal hormones begins early in development and perhaps before morphological differentiation suggests that these hormones play an important part in early gonadal differentiation

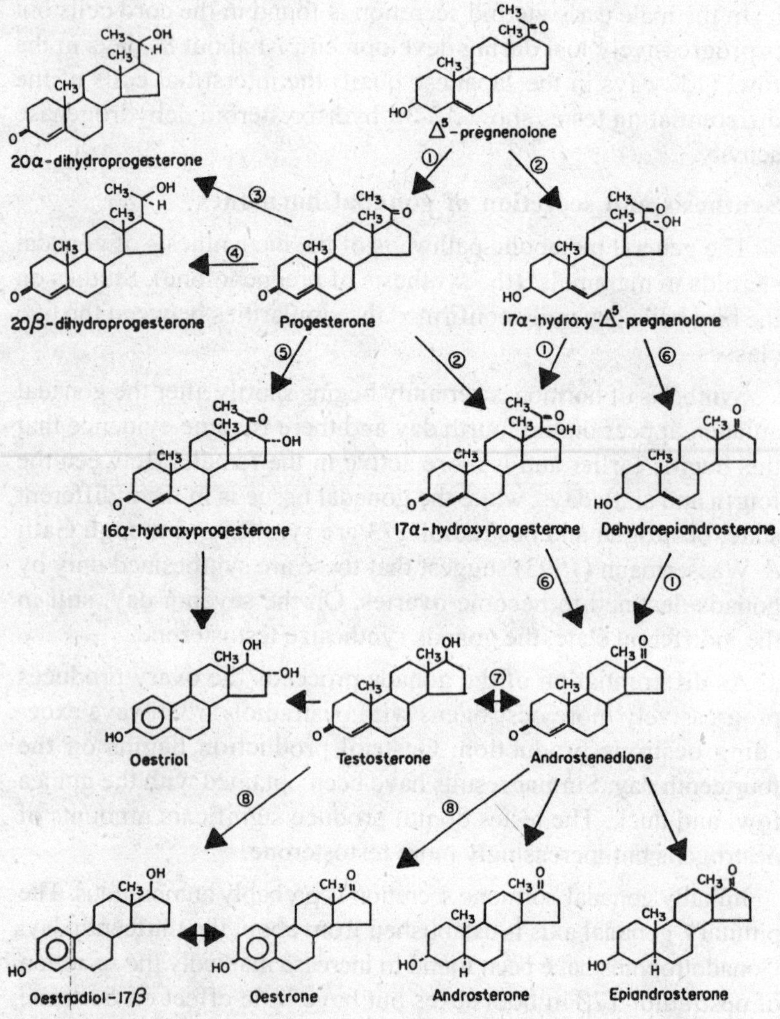

Figure 10.9 : Metabolic pathways for the synthesis of gonadal steroid hormones. The enzymes involved are as follows

(1) ΔS-3β-ol-dehydrogenase and Δ 5-isomerases
(2) C_{21} steroid 17α-hydroxylase
(3) 20α−dehydroxy-dehydrogenase
(4) 20α-dehydroxy-dehydrogenase
(5) 16α-hydroxylase
(6) 17-20-desmolase
(7) 17β-hydroxysteroid dehydrogenase
(8) Aromatizing enzymes

and development. The finding of earlier and greater activity by the gonad destined to become ovarian in nature conforms with this notion. However, with perhaps two exceptions, the roles of the various hormones remain to be properly elucidated.

The mechanism of regression of the Miillerian ducts differs between sexes. Both regress in the male but only the right one in the female. In the male regression is controlled by the testicular secretions. The active principle is not testosterone but remains to be positively identified though there is some evidence that it may be an oestrogen-like substance. It is known that the maximum activity of the 'regression' hormone is from 11-13 days and it is suggested that the subsequent fall in activity is related to the onset of the secretion of thyrotrophin, follicle-stimulating hormone and growth hormone by the pituitary.

The regression of only the right Miillerian duct in the female is probably stimulated by the oestrogens though it is likely that there is also a genetic component at work. Studies on the hatched bird have indicated that oestrogen is important in stimulating development and that progesterone generally accentuates this action. Thus it is the right duct that regresses even in embryos where the left ovary has been ablated.

The androgens, particularly testosterone, are required for the transformation of the Wolffian ducts into the vasa deferentia.

Some instances of the gonadal hormones influencing an extra-gonadal metabolic pathway have been tentatively identified. Young embryos have been found to be able to degrade testosterone only to 5p-reduced metabolites. These 5p-androstanes induce 8-aminolevulinic acid synthetase, the rate-limiting enzyme in porphyrin synthesis. Thus the embryo may well be able to influence erythro-poietic activity. There is also some evidence that testosterone may stimulate chondroitin sulphate synthesis in bone while oestradiol-17p may reduce it. There are, furthermore, the necessary enzymes present in the embryonic cartilage to catabolize the sex hormones. Finally, and perhaps not unexpectedly in view of the marked sexual dimorphism in the plumage of sexually active birds, the gonadal hormones affect the pigmentation of the down feathers.

Brooks & Ungar (1997) have reported that progesterone, 16- or 17hydroxyprogesterone stimulates the maturation of the so-called 'hatching muscle' *(musculus complexus)* which is concerned in the pipping and hatching processes. Progesterone not only stimulated the uptake of water, a process that is probably unique to this muscle, but also advanced hatching. Progesterone might therefore be considered for the role of a hatching hormone. However, attempts to repeat this work have not been successful.

HYPOPHYSEAL HORMONES

The pituitary gland or the hypophysis consists of two distinct parts, the adenohypophysis and the neurohypophysis: further subdivisions are possible. Its endocrine functions arc of two basic types: the control of the activity of other endocrine glands, such

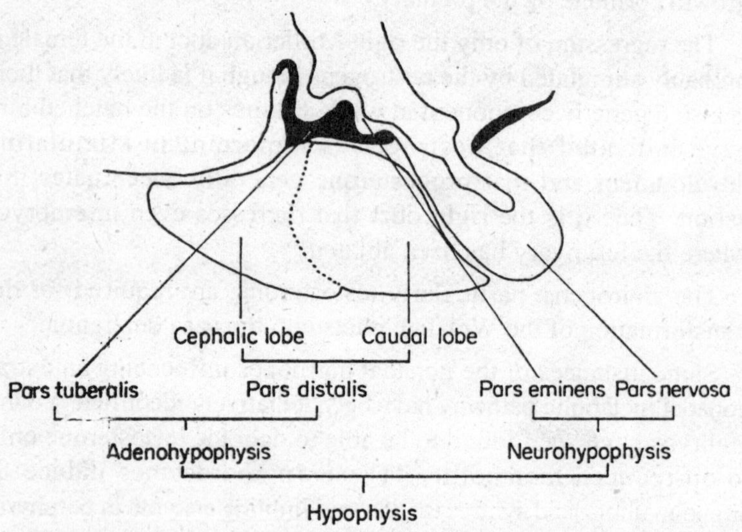

Figure 10.10 : A generalized scheme of the avian hypophysis showing the main divisions. Note that the pars tuberalis (solid areas) forms a collar around the infundibular stem. There is no pars intermedia.

as the thyroids, adrenals and the gonads and the production of hormones which have a general metabolic effect. But while the hypophysis may be seen as the co-ordinator of other endocrine glands it is itself, at least in the hatched bird, influenced by higher centres of activity, notably the hypothalamus. However, as we shall see its activities in the embryo are independent of hypothalamic control.

Development of the hypophysis

The adenohypophysis is derived from the oral epithelium of Rathke's pouch which extends upwards, from the second day (stage 10), towards the developing infundibulum. Concurrently there is an outgrowing from the infundibulum to form the neurohypophysis. Together these two diverticuli become associated to form the hypophysis. Further details of their embryological development and cytodifferentiation may be found in Wingstrand (1981), Romanoff (1990), Doskocil (1995, 1996) and Mikami *et al.* (1993).

The adenohypophysis is made up of the pars distalis and the pars tuberalis : there is no pars intermedia in the birds as there is in mammals. The pars distalis, perhaps the most studied portion of the hypophysis, is usually subdivided further into the cephalic and caudal lobes. The cephalic lobe probably secretes adrenocorticotrophic hormone (ACTH), thyroidstimulating hormone (thyrotrophin: TSH) and follicle-stimulating hormone (FSH) exclusively; the caudal lobe probably secretes lute' nizing hormone (LH: also known as interstitial cell-stimulating hormone, ICSH) while growth hormone (somatotrophin: GH) may be produced by either only the caudal lobeor both lobes. It is uncertain whether melanocyte-stimulating hormone (MSH) or prolactin are secreted by the pars distalis. Species differences may well occur, however. TSH may be secreted by both lobes in the duck, the pigeon, by only the cephalic lobe in the white-crowned sparrow and the Japanese quail.

Little is known of the pars tuberalis. It usually covers part of the diencephalon and encircles the infundibular stem.

The neurohypophysis remains distinct from the adenohypophysis, there being a sheath of connective tissue between them. This neural lobe, the equivalent of the posterior lobe of the mammal, contains neurosecretory fibres which extend back to the median eminence (pars eminens) and the hypothalamus. Neurosecretory material becomes visible in the neurohypophysis on about the thirteenth day. Thereafter the amount of material increases rapidly.

Secretion of hormones by the pars distalis

As we have already noted little or no information on the secretions of the pars tuberalis of the adenohypophysis or the neurohypophysis is available. We shall therefore confine this section to a discussion of the hormones of the pars distalis of the adenohypophysis.

It is clear from the literature that the various cell-types of the pars distalis mature at different but precise times in development. ACTH is generally considered to be the first trophic hormone to be secreted: in the domestic fowl it probably appears on the eighth day. On the tenth day TSH secretion begins; on the thirteenth day the secretion of gonadotrophins (LH and FSH) and on the fifteenth day the secretion of GH.

Significance of the hypophyseal hormones in development

Hypophyseal hormones are essential for normal development. Only some 6% or so of hypophysectomized embryo survive to 21 days and of these none hatches whereas hypophysectomized embryos with a par distalis grafted on to their chorio-allantoic membrane show an almost complete restoration of their viability and physiology. It is therefore not surprising that hypophysectomized embryos are dwarfed (lack of GH), have small, non-functioning thyroids (lack of TSH), adrenals (lack of ACTH) and gonads (lack of gonadotrophins). It is perhaps a little surprising, however, that all these hormones are secreted by the pars distalis acting autonomously, that is without the need for the so-called 'releasing factors', thyrotrophin releasing factor, adrenocorticotrophin releasing factor, etc., secreted by the hypothalamus. The successful restoration of a normal physiological state by grafting the pars distalis on to the chorio-allantoic membrane of the hypophysectomized embryo confirms this to be the case.

The evidence for a role by GH in the differentiation of the duodenum remains equivocal: Bellware & Betz (1990) considered it necessary but more recent work by Hart & Betz (1992) has thrown doubt on this. Betz (1997) found that the extra-embryonic membranes are poorly developed in the hypophysectomized embryo and suggested that a hormone from the pars distalis stimulates their development. It is conjectural that this hormone might be GH.

The progressive inhibition of the Miillerian duct regression hormone is probably accomplished by the increasing activity of FSH, TSH and GH. On the other hand, the hypophyseal gonadotrophins are not required for the transformation of the female left Miillerian duct into the oviduct.

It has recently been suggested that the pituitary might influence directly bone differentiation and mineralization. The evidence is admitted to be slender in view of the known effects of the trophic hormones on other endocrine glands which in turn affect bone metabolism and much more work will be needed before a direct role can be assigned to the adenohypophysis.

Several authors have noted that the hypophysectomized embryos suffer from a disturbed water balance. This disturbance can be partially relieved by a pars distalis graft, perhaps through a restored secretion of aldosterone, though it seems reasonable to assume that the neurohypophysis plays some part in this aspect of the physiology of the embryo.

GLANDS OF UNCERTAIN ENDOCRINE STATUS

In the preceding sections of this chapter we have considered the known endocrine glands in some detail. There are, in addition, several glands whose endocrinological status is uncertain. Inevitably the evidence for considering some of these as endocrine glands is stronger than for others. Thus the kidney (mesonephros, metanephros) and liver may secrete erythropoietin - the thymus may secrete a hormone concerned in the maturation of the lymphoid system; the bursa of Fabricius may secrete a hormone with similar activity to that of the thymus and the pineal gland may produce a hormone which acts upon the hypophysis to enhance growth and maturation of the embryo. In this section we shall consider two of these: the bursa of Fabricius and the pineal gland.

Bursa of Fabricius

This gland arises as a diverticulum of the proctodaeum of the cloaca. Its function has yet to be precisely defined though there is now sufficient evidence to sustain the concept of its being endocrine. The evidence that it produces secretions concerned with immuno logical competence is particularly strong. Thus in view of Thorbecke *et al.* (1998) secretion of a bursal hormone begins in the period 16-18 days, the hormone being transported either from cell to cell or by the migration of hormone-producing cells to other peripheral lymphoid sites. They consider that the hormone stimulates differentiation and maturation of the lymphocytes.

The pineal gland

The demonstration that embryos exposed to light during incubation tend to hatch earlier than those incubated in total darkness indicates that the embryo is photosensitive. There are two known discrete photosensitive organs in the bird, the eye and the pineal gland or epiphysis cerebri.

At present it is not known which if either of these two organs might be concerned with the reception of the photostimulus and its transmission to the hypophysis though the late maturation of the eye makes it a doubtful candidate. Possible functions of the pineal gland have been fairly exhaustively explored by Stalsberg (1995) who found that the incubation period was unaffected by pinealectomy *in ovo* though regrettably the influence of light *per se* was not studied. The involvement of the pineal gland in determining the rate of development thus remains conjectural.

11

Physiology of Hatching

Terminating the embryonic existence, a process usually described by the deceptively simple word 'hatching', is in fact an extremely complex procedure and it is not surprising, therefore, that there is a peak of mortality at this time. Unlike the mammal, essentially all the stimuli are generated by the embryo itself, though it is able to respond to stimuli from both the parents and the other members of the clutch.

Superficially there are three major events in the sequence of hatching: the onset of pulmonary respiration, pipping (that is the single point fracture of the shell, usually in that overlying the air space), and the emergence of the hatchling. However, underlying these more obvious events are many others, e.g. maturation of muscles, changes in the circulation, withdrawal of the yolk sac, all following a sequence which will ensure a successful outcome of incubation.

PULMONARY RESPIRATION

Birds and reptiles pass through a stage of development unknown to the mammal - a period when the embryonic and adult respiratory surfaces, the chorio-allantois and the lungs, function side by side. This period is conveniently described as the parafetal period and the animal, the parafetus. These terms are due to Romijn (1948) and were originally used to describe the period from the onset of breathing to pipping. Here we shall use the term to cover the period when the two respiratory systems function simultaneously.

The respiratory system

The embryology of the system is fully described by Hamilton (1982) and Romanoff (1990). In brief the system develops from two distinct sites: the larynx and trachea are derived from the median laryngotracheal grooves; the lungs, bronchi and air sacs develop from paired endodermal diverticula of the embryonic fore-gut. The air sacs, which appear on the sixth day, are initially found as six paired structures but two pairs fuse to form the median clavicular sac and, in some species including the fowl, another pair fuse to form the median cervical sac. Thus there are normally eight or nine sacs : the median (or paired) cervical sac(s), the median clavicular and the paired cranial thoracic, caudal thoracic and abdominal sacs. These sacs are connected, via the ostia, to the bronchi and are concerned, in the hatched bird, with the movement of air within the system. Their innervation, in the fowl and mute swan, is described by Groth (1992).

The architecture of the bronchi is complex. Each lung is supplied with one primary bronchus (the mesobronchus) which gives rise to the secondary bronchi (ento-, ecto-, latero- and dorsobronchi) and these in turn give rise to the tertiary bronchi (parabronchi). No tertiary bronchus ends blindly and manyy join to form long, curved circuits. It is these tertiary bronchi that are concerned with gaseous exchange. Each is pierced by numerous openings which lead in turn to the atria and then the infundibula. The infundibulum leads to a network of fine anastomosing air capillaries which are in intimate contact with the blood capillaries. Further details of the structure of the lung of the hatched bird may be found in King & Molony (1991).

The flow of air within the respiratory system is complex and the air sacs, far from being evolutionary relics or buoyancy bags, have an important function in maintaining the correct pattern of air flow. They are not concerned with gaseous exchange *per se,* however. There is no true diaphragm. The pattern of air flow is fully discussed by Brackenbury (1991, 1992a, b), Bouvert & Dejours (1991), Bretz & Schmidt-Nielsen (1991, 1992) and Scheid & Piiper (1991). It will be sufficient to point out here that both inspiration and expiration are active processes and, because of the flow pattern, fresh air is probably passing unidirectionally through the tertiary bronchi continuously. The avian respiratory system is, then, an extremely specialized system, designed to meet the heavy energy requirements of flight, and, as a result, is able to supply the bird with more oxygen per unit time than any other system.

The onset of breathing: the pulmonary stimulus

A prerequisite for breathing is the removal of the fluid invading the whole of the respiratory tract. Undoubtedly a major component of this fluid is of amniotic origin, but within the lung itself the fluid may contain a large proportion of an ultrafiltrate of the embryo's blood. The uptake of amniotic fluid, by active imbibition, is begun, in all birds examined, when 70% of the incubation period is complete and is virtually complete just prior to the initiation of breathing. Thus in the fowl imbibition is completed about a day before the chick hatches. While some of the fluid remaining in the respiratory tract proper may flow into the pharynx and then be swallowed, some remains and presumably is taken up by the lung tissue itself.

Early attempts to identify the pulmonary stimulus, though not completely successful, served to indicate that it is probably gaseous and that it is possibly a high partial pressure of carbon dioxide in the arterial blood $\left(Pa_{CO_2}\right)$. Thus Windle & Barcroft (1988) with the fowl, and Windle & Nelson (1988) with the duck, found that clamping the umbilical vessels of the mature embryo stimulated deep rhythmic respiratory movements almost immediately. That carbon dioxide was the more likely stimulus was suggested by the observation that respiratory movements could be initiated by raising the P_{co_2} in the atmosphere surrounding the egg by only 12 mmHg while a reduction of 68 mmHg in P_{O_2} was required.

It is only comparatively recently that there has been further interest shown in the nature of the pulmonary stimulus. Visschedijk (1982, 1988b) found that increasing or decreasing the permeability of the shell over the air space did not lead to any change in the timing of pulmonary respiration. However, Freeman (1984a) using paraffin wax instead of liquid paraffin for reducing permeability found a significant advance of five hours. He confirmed that perforating the shell was without effect as was ventilating the air space.

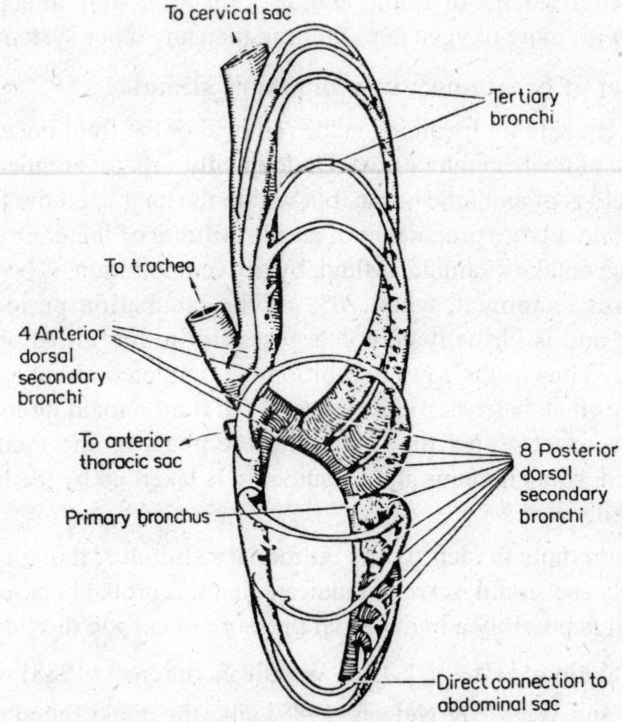

Figure 11.1 : Bronchi of the goose lung. Dorsal view of the right lung; the ventral series of posterior secondary bronchi has been omitted.

At first sight it would seem that either the enhanced P_{CO_2}, reduced P_{O_2} or both in the air space (Visschedijk, 1988c) was responsible for stimulating breathing. There is an important objection to this interpretation, however: perforating or ventilating the air space failed to delay breathing.

Another approach to the problem has been made possible by the recent development of micro-techniques suitable for analysis of gases in the small volumes of blood available from the avian embryo. Towards the end of the embryonic period the $PaCO_2$ may reach the very high level of 60 mmHg (Dawes & Simkiss, 1989; Freeman & Misson, 1990). PaO_2 is relatively low - about 20 mmHg - but not exceptionally so. The likelihood of a high $PaCO_2$ being the stimulus is strengthened therefore by these direct observations, though the further observation that there is a significant fall in $PaCO_2$ before breathing commences weakens the hypothesis.

In an attempt to resolve this paradox we have examined a modification of the hypothesis suggested by the above results - that the stimulus is a particular ratio of $PaCO_2$ to PaO_2 as it is in the sheep - but without success.

A different approach has been suggested by the demonstration that all aspects of the hatching process can be substantially advanced by sound - 'clicking'. This suggests that under certain circumstances the gaseous stimulus can either be initiated earlier or overriden by the sound stimulus. An earlier initiation of pulmonary respiration has been observed in the fowl, duck, goose and Japanese quail but not in the bobwhite quail.

The technique of sound stimulation has been used in conjunction with blood gas analysis in an attempt to determine whether or not there is a unique change in the partial pressures of the blood gases prior to breathing. Such a change might be interpreted as the pulmonary stimulus. However, while sound stimulation led to an advance in the initiation of breathing no clear differences in $PaCO_2$ or PaO_2 could be identified.

The positive identification of the pulmonary stimulus remains an intractable problem therefore and it seems very likely that it is gaseous but whether it is a high $PaCO_2$, a low PaO_2 or a critical ratio of the two still remains to be elucidated.

The development of the breathing pattern

Breathing probably begins about the time the beak penetrates the inner shell membrane though the embryo is capable of showing respiratory movements many hours before. The lungs neither inflate immediately nor completely. Rather the tertiary bronchi are aerated first in limited areas. These areas can be recognized by the change in colour of the lung from dark to bright red. M. A. Vince and B. Tollhurst (personal communication) have found that inflation is complete in both lungs after about 6 h (the fowl), 4 h (duck) or 9 h (bobwhite quail). The role of surfactant in facilitating breathing has not been completely elucidated.

Initially breathing is both irregular and intermittent. Gradually regular patterns of medium amplitude breathing emerge and come to dominate the pattern. In the third phase both frequency and amplitude increase and each breath is often accompanied by a 'click sound'. This pattern has been observed in the several species examined though the length of any one phase may vary from species to species.

The rate of breathing in the hatching fowl reaches about 90 or 100 breaths per minute while in the duck it is a little higher at 110 breaths per minute.

Recent work by Dawes (1993) indicates that the control mechan-isms of respiration arc at lcast relatively mature at hatching. Thus during experimental anoxia he found that breathing amplitude rose quickly indicating some peripheral stimulation. Thereafter the parafetus began to gasp, breathing rate slowed and finally ceased illustrating the excitatory and depressant effects respectively of anoxia on the respiratory centres of the brain.

Circulatory and associated changes

The circulation of the embryo is characterized by two short circuits — the ductus arteriosus between the pulmonary and aortic arches and the intraatrial foramina, the avian equivalent of the foramen ovale. The ductus venosus, unlike its mammalian counterpart, is lost on the seventh day of incubation. When the bird commences breathing these short circuits, together with the chorio-allantoic circulation, have to be closed off in order that a double circulation can be formed.

It has been found that the walls of the ductus arteriosus have well developed muscle layers, particularly at the proximal end. As breathing begins these muscles contract to restrict and finally prevent blood flow (Harms, 1997). This is a gradual process, and is complete in 91% of the chicks at hatching (Coughlin & Husson, 1990). The physiological mechanism is uncertain, neither acetycholine nor catecholamines have marked stimulatory properties on the muscle cells.

The intra-atrial foramina are so constructed that they normally act as at their great valves. During diastole of the right atrium the blood flows into the chamber Blood pres and pushes the flaccid inter-atrial septum towards the left thereby opening and continue the formina and allowing a proportion of the blood to flow directly into the left atrium.

During systole the pressures in the atria become equalized, thus allowing the septum to take a medial position and in so doing the foramina are closed. As breathing becomes established more blood flows into the left atrium from the now-functioning pulmonary vein and as a result the pressures within the atria tend to be equal. This closes the foramina and their permanent closure follows, usually within four or five days.

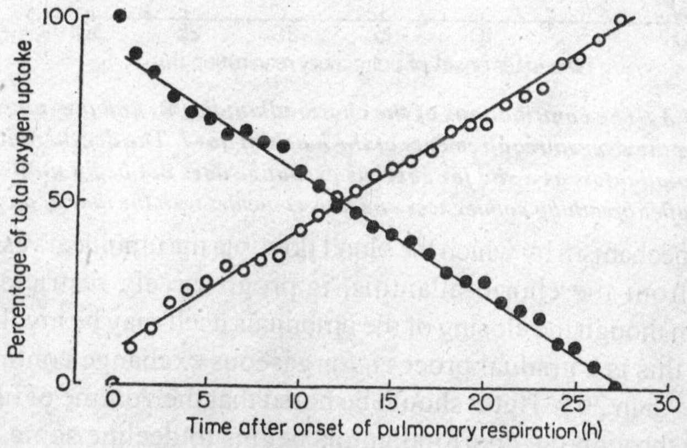

Figure 11.2 : The percentage contributions of the chorio-allantois (●) and the lungs (O) to the oxygen requirements of the hatching chick. The data for this figure were calculated from Visschedijk (1962). The embryo pipped the shell after 10 hours and hatched 27 hours after the commencement of breathing.

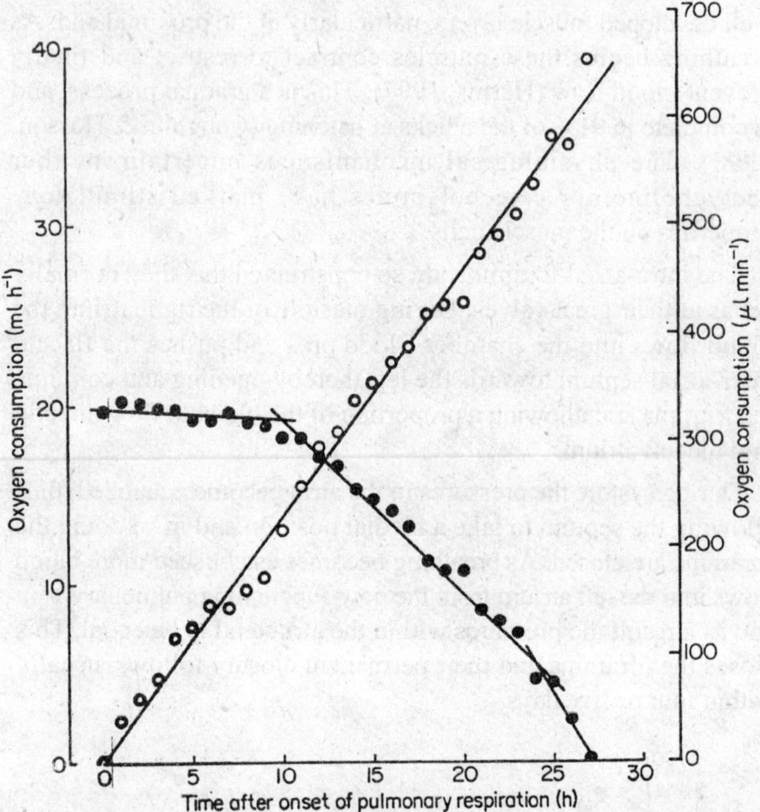

Figure 11.3 : The contributions of the chorio-allantois (●) and the lungs (O) to meeting the oxygen requirements of the hatching fowl. The degeneration of the chorio-allantois as a site for gaseous exchange does not begin until some 10 hours after breathing commences and thus coincides with the time of pipping.

The mechanism by which the blood flow, via the umbilical vessels, to and from the chorio-allantois, is progressively restricted is uncertain though the closing of the umbilicus itself may be involved. Clearly this is a gradual process, for gaseous exchange continues for more than 20 h. But it should be noted that the volume of blood flowing through the chorio-allantois begins to decline some four days or so earlier. Why this is so is uncertain and in some ways paradoxical. It is precisely at this period of development that the problems of oxygenating the embryo are probably at their greatest.

Blood pressure rises markedly during hatching to at least 43/23 mmHg and continues to increase in the immediate post-hatching period and the heart rate increases from about 260 beats per minute to 295 beats per minute or more.

Pipping

Some 8 or 9 h after breathing is established in the fowl the shell over the air space is usually fractured at one point - 'pipped'. The timing of this event can be advanced by waxing or oiling this part of the shell, delayed by perforating it or virtually abolished by ventilating the air space with atmos pheric air.

Thus pipping, like breathing, would seem to result from a gaseous stimulus. In a series of elegant experiments Visschedijk (1968c) has established that either a reduction in the P_{O_2} or a rise in the P_{CO_2} in the air space advances the time of pipping and that for a given change, carbon dioxide is twice as effective as oxygen.

Just prior to pipping the P_{O_2} in the air space of the domestic fowl may fall to as little at 60 mmHg while the P_{CO_2} may reach 60 mmHg. At this point in the hatching process the lungs are meeting about 40% of the total oxygen requirements of the parafetus and it might be expected that the parafetus will be faced with increasingly hypoxic conditions. However, there is little evidence that this occurs : cardiac stores of glycogen are normally low, suggesting few problems during hatching though a transient depletion can be detected at pipping. There is, furthermore, no significant accumulation of lactate and the Pa_{O_2} and Pa_{CO_2} do not reflect hypoxia.

General activity, including that of the 'hatching' muscle *(musculus complexus)*, increases as the parafetus moves further into the air space. Eventually the beak comes into contact with the calcareous shell and the enhanced muscular activity leads to the egg tooth - the reinforced tip of the upper beak - penetrating it. Once the shell is pipped the parafetus enters a more quiescent period, at least in terms of physical activity, which persists until 3 or 4 h before the chick emerges from the shell.

ACTIVE HATCHING

The final activity of the parafetus is directed towards cutting round the shell and the emergence of the chick. That this period is independently controlled was first suggested by Freeman (1992). He found that the oxygen consumption of the parafetus began to rise an hour or two before active hatching began. It was suggested that this increase in metabolism resulted from the release of a hormonal stimulus which leads to the active cutting round of the shell.

It is now generally agreed that the hatching stimulus is not gaseous. Visschedijk (1992, 1998b) and Freeman (1994a) found that altering the permeability of the air space shell did not have any effect on the time of hatching. The suggestion that it is hormonal in nature was strengthened by the demonstration that the thyroid hormones accelerate hatching and that they stimulate active hatching specifically. However, it has been suggested that progesterone is the hormone responsible, though Oppenheim (1993) has been unable to repeat their experiments.

An important role has been ascribed to the muscle, *musculus complexus,* in the hatching processes. This muscle shows marked development just before hatching begins and acts to raise the head of the parafetus and is presumed to provide most of the force necessary for breaking down the shell. The timing of its maturation seems to be unique but attempts to explain this have largely failed.

WITHDRAWAL AND FATE OF THE YOLK SAC

As incubation draws to a close there remains outside the embryo a variable amount of yolk in the yolk sac. This is progressively withdrawn into the abdominal cavity, from the nineteenth day in the fowl, probably as a result of the activity of the abdominal musculature, the process in the fowl being completed some 14 h before emerging from the shell. The control of the withdrawal activity is uncertain, though there is limited evidence that. the thyroid or adrenal or both may be involved.

Although the yolk sac is connected directly with the duodenum there is little or no movement of material via this route even after hatching. Instead it is absorbed through the yolk sac membrane and transported to the chick by the omphalomesenteric vessels. There are approximately 5 g of yolk in the yolk sac of the neonate fowl but this is rapidly utilized by the bird and the yolk sac is usually vestigial by the fifth day.

OXYGENATION AND ENERGY METABOLISM DURING HATCHING

With the onset of breathing the oxygen requirements of the parafetus are met by the combined activities of the chorio-allantois and lungs. It is possible to determine their relative contributions during the parafetal period from the data of Visschedijk (1992).

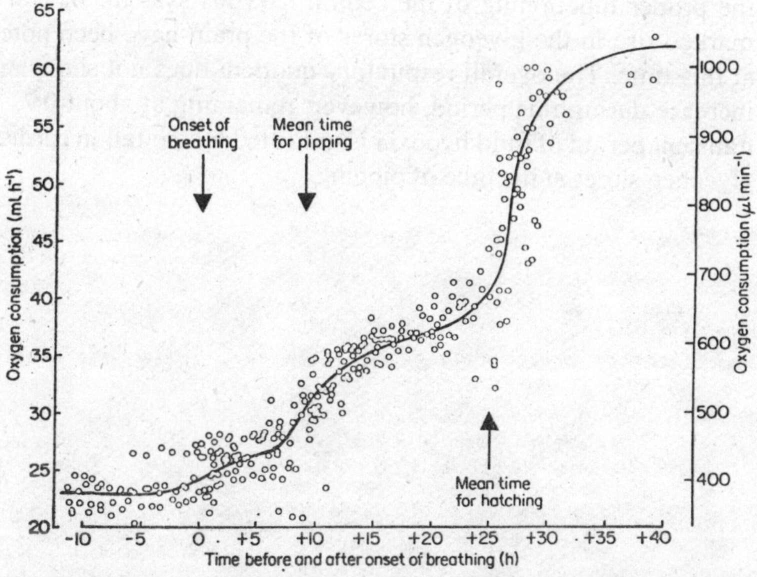

Figure 11.4 : Oxygen consumption of the fowl during the hatching period. The data have been taken from Visschedijk (1992) and Freeman.

Given that the reduction in the chorio-allantoic component is similar over its whole surface, we can see that after the initial 3 h or so the change-over from chorioallantoic to pulmonary respiration is linear and is not completed, at least in the fowl, until the very end of the hatching period. However, the degeneration of the chorio-allantois would not seem to be initiated until 10 h after the initiation of breathing for it continues to provide the same amount of oxygen up to that point. Thereafter its importance as a respiratory surface declines rapidly.

The oxygen requirements of the parafetus, not unexpectedly, rise during hatching. Typically the fowl consumes 25 ml h^{-1} prior to breathing, 35 ml h^{-1} prior to active hatching and about 40 ml h^{-1} at the moment of escape from the shell. After this, at least in the homeothermic species, a further substantial rise occurs.

It would seem that the bird's requirement for carbohydrate increases markedly during hatching for there is a dramatic mobilization of glycogen stores a little before breathing commences. It is likely that the glucose that is made available is necessary for the proper functioning of the central nervous system. Indeed a marked rise in the glycogen stores of the brain have been noted at this time. The overall respiratory quotient does not show any increase during this period, however, remaining at about 0·7. A transient period of mild hypoxia is indicated by the fall in cardiac glycogen stores at the time of pipping.

12

The Neonate

Birds show a wide range in their degree of functional maturity at hatching. Some species, the duck for instance, are relatively mature, in that they have sight, are able to feed themselves and generally show a marked degree of independence. Others, the passerines for instance, are distinctly immature being blind and naked and totally dependent upon the parent birds for their survival. But whether the hatchling is mature or immature all neonates are confronted by the same hostile environment. In this chapter we shall discuss three of the many problems : changing environmental temperature, digestion and absorption of food and the problem of bacterial and viral diseases.

THERMOREGULATION

Ontogeny of thermoregulation

Birds and mammals are the only two classes able to produce heat for the maintenance of their body temperature. They may therefore be termed endothermic homeotherms and are thus distinguishable from the ectothermic homeotherms which, while able to control their body temperature, do so by deriving heat from outside the body.

While the adults of the two classes are homeotherms their embryos are poikilothermic. Homeothermy, furthermore, does not necessarily develop at birth: when it does the species is said to be precocious. Those species which become homeotherms later are described as altricial. This concept, while useful, is too rigid and a complete range of types may be distinguished. While they may show marked metabolic responses to cold immediately after hatching not all the precocious species are complete homeotherms even within a day of hatching. Of those species studied the Anatidae are probably the best able to maintain their body temperature; similarly the slender-billed shearwater and the western gull are quite good while the Galliformes are less good and the Japanese quail is very poor. Most precocious species have very good thermoregulatory ability by the end of the first week of post-embryonic life.

The ability to regulate heat production in the altricial species develops a variable, though constant for each species, time after hatching and may not be complete until the end of the nestling period.

Mechanisms of thermoregulation

The attainment of homeothermy in both altricial and precocious species has often been correlated with the development of insulation, in particular the feathering (including the down feathers). There is no doubt that this is a factor but it is not the most important one. Perhaps the rapid increase in the ratio of mass to surface area is more important. Freeman (1995b) has shown that part of this increase in mass is brought about with no change in surface area by the replacement of the relatively inert yolk by metabolically active tissue. At the same time Wekstein & Zolman (1988, 1989, 1990) have shown that the capacity to thermoregulate depends on age rather than the degree of feathering.

Chemical regulation of heat production has been studied mostly in the fowl. While the neonate is capable of a limited amount of shivering it seems likely that a proportion of thermoregulatory heat is produced by non-shivering means though not through the activity of brown adipose tissue for the fowl at least does not possess this tissue. The thermogenic stimulant may well be thyroxine. The ability to shiver develops rapidly after hatching and is probably responsible for most of the thermoregulatory heat produced after a week. A similar situation exists in other avian species both precocial and altricial.

Precocious species are able to pant or utilize gular fluttering but the altricial species show little ability to increase heat loss by these methods.

Most neonate birds, whether altricial or precocious show behavioural responses to temperature. They will huddle and reduce their activity to a minimum in order to conserve energy. Klciber & Winchester (1983) have shown that such actions lead to a 15% reduction in energy expenditure in the domestic fowl.

THE ALIMENTARY TRACT

In general the smaller the bird the smaller, relatively, are its food reserves at hatching. During incubation the domestic fowl utilizes 42% of its lipid stores whereas the eastern house wren uses 99%. It is not surprising therefore that the alimentary tract is usually mature at hatching. However, as we shall see, some digestion probably occurs before hatching.

The maturation of the tract has been best studied in the domestic fowl where it has been shown that there is a marked increase in development during the final 6 days or so of incubation. Thus the oesophageal mucus glands are mature at hatching; the gland cells of the proventriculus commence secretion on the thirteenth day; the innervation of the gizzard becomes operative at hatching; the gizzard itself is mature at hatching; the duodenum and small intestine enter their final phase of differentiation three or four days before hatching; the gut transport systems for amino acids and sugars become functional, and for the latter become markedly more efficient; the activities of pancreatic chymotrypsin, carboxypeptidase, amylase, maltase, sucrase and lipase increase dramatically from the seventeenth day. The changes in invertase activity do not parallel the changes shown by these enzymes, however. Activity increases progressively from 13 days, falls significantly at 20 days and then increases again.

A limited amount of digestion begins before hatching for the embryo actively imbibes the amniotic fluid from the thirteenth day.

This fluid is particularly rich in protein for the albumen floods through the seroamniotic connection into the amniotic cavity from the twelfth day. It is of interest, therefore, that the amino acid transport system of the small intestine should be mature by the sixteenth day. Recent work has shown that the major molecular species of chymotrypsin in the duodenum at this time is chymotrypsin 3. Since this species is rapidly replaced by chymotrypsins 1 and 2 after hatching Cohen and Kulka suggest that this species is particularly active towards the albumen proteins, one of the most prevalent amino acids being leucine. At the same time, however, it seems likely that intact protein can be transported across the gut epithelium for antibodies of maternal origin can be detected in the embryo. This ability is lost at hatching.

Finally it should be pointed out that the residual yolk is not transferred through the yolk stalk directly into the small intestine for digestion. Fritz (1991) has shown that the yolk material is absorbed exclusively via the omphalo-mesenteric blood vessels.

IMMUNOLOGICAL COMPETENCE

It is generally assumed that the avian embryo develops in a sterile environment. By and large this is true but it is not always the case for some diseases, for instance, infectious avian encephalomyelitis, are transmitted from the mother to her progeny via the egg. The cuticle of the egg is an important barrier to the penetration of the shell by bacteria during incubation and even should some organisms pass through the pores into the egg the anti-bacterial properties of the albumen often prevent the establishment of an infection. Once the shell is fractured during the hatching process the bird becomes exposed to a contaminated environment and the survival of the bird will largely depend upon its ability to react immunologically too the antigenic stimuli.

The neonate usually possesses a variety of circulating antibodies but all these have been absorbed, probably intact, from the yolk, having been secreted by the hen. Thus the chick is passively protected for up to four weeks after hatching by these maternal antibodies. The ability of the neonate to respond to novel antigenic stimuli has been shown to exist at hatching, though not before and then only to a very limited degree. This primary antigenic response develops rapidly and progressively after hatching and is mature after three weeks.

The importance of the bursa of Fabricius in conferring immunological competence upon the chick is now recognized.

CONCLUDING REMARKS

In the life-cycle of an animal any stage is the product of earlier ones and at the same time it must necessarily influence future stages. Thus the embryo is affected by the availability of nutrients laid down by the hen, the newly-hatched chick is the product of incubation and the adult is, in turn, the product of that newly-hatched chick as further influenced by the environment.

We have, of course, dealt here with only part of the life-cycle - that part conveniently identified with the egg. It will be clear from both sections of the book that the concept of the cleidoic egg being complete and isolated is too simple. The embryo is not only affected by many environmental factors but also reacts to them. Thus there is a constant interaction between embryo and environment.

Hatching should not therefore be considered as an end-point for it is simultaneously a beginning. It is perhaps nothing more than a convenient reference point.

Index

O

P

R

S